RESTRICTED

TM 9-750

ORDNANCE MAINTENANCE

LEE MEDIUM TANKS M3, M3A1, AND M3A2

TECHNICAL MANUAL

May 9, 1942

by WAR DEPARTMENT

©2013 Periscope Film LLC
All Rights Reserved
ISBN#978-1-937684-35-8
www.PeriscopeFilm.com

DISCLAIMER:

This manual is sold for historic research purposes only, as an entertainment. It contains obsolete information and is not intended to be used as part of an actual operation or maintenance training program. No book can substitute for proper training by an authorized instructor.

©2013 Periscope Film LLC
All Rights Reserved
ISBN#978-1-937684-35-8
www.PeriscopeFilm.com

IT TAKES *care*

TO KEEP AN ENGINE RUNNING

During recent army maneuvers most engine failure was due to improper care.

To keep your tank running, be absolutely sure to:

- ✓ Idle the engine always at minimum of 800 rpm *except* when shifting gears.

- ✓ Change engine oil at 25-hour intervals when operating in dust; never longer than 100 hours under ideal conditions.

- ✓ Clean air cleaners *daily*.

- ✓ Turn handle of air filter *at least* once daily. Drain at 25 hours.

- ✓ Warm engine to at least 80° F. before moving tank.

- ✓ Stop tank if oil gets hotter than 190° F.

*TM 9-750

RESTRICTED

TECHNICAL MANUAL }
No. 9-750

WAR DEPARTMENT,
Washington, May 9, 1942.

ORDNANCE MAINTENANCE

MEDIUM TANKS M3, M3A1, AND M3A2

	Paragraphs
CHAPTER 1. Operating instructions.	
SECTION I. General	1-3
II. Description and tabulated data	4-5
III. Operation instructions and controls	6-14
IV. Lubrication	15-18
V. Inspections	19-24
VI. General care and preservation	25-32
VII. Matériel affected by gas	33-34
VIII. Armament	35-37
CHAPTER 2. Organization instructions.	
SECTION I. General information on maintenance	38-39
II. Equipment and special tools	40-41
III. Engine and accessories	42-60
IV. Fuel system	61-67
V. Cooling system	68-69
VI. Clutch	70-81
VII. Propeller shaft	82-84
VIII. Transmission, differential, and steering brakes	85-90
IX. Final drive	91-93
X. Tracks and suspensions	94-99
XI. Electrical equipment and instruments	100-118
XII. Auxiliary generating unit	119-121
XIII. Stabilizers	122-126
XIV. Turret	127-131
XV. Preparation for shipment and storage	132-134
	Page
APPENDIX. List of references	154
INDEX	155

*This manual supersedes TM 9-750, October 1, 1941.

CHAPTER 1

OPERATING INSTRUCTIONS

	Paragraphs
SECTION I. General	1–3
II. Description and tabulated data	4–5
III. Operation instructions and controls	6–14
IV. Lubrication	15–18
V. Inspections	19–24
VI. General care and preservation	25–32
VII. Matériel affected by gas	33–34
VIII. Armament	35–37

SECTION I

GENERAL

	Paragraph
Purpose and scope	1
Content and arrangement	2
References	3

1. Purpose and scope.—These instructions are published for the information and guidance of the personnel of the using arms charged with the operation, maintenance, and repair of this matériel. They contain descriptions of the major units and their functions in relation to the other components of the tank, as well as instructions for operation, inspection, minor repair, and unit replacement.

2. Content and arrangement.—Chapter 1 contains information chiefly for the guidance of operating personnel. Chapter 2 contains information intended chiefly for the guidance of personnel of the using arms doing maintenance work.

3. References.—Appendix I lists all Standard Nomenclature Lists, Technical Manuals, and other publications for the matériel described herein.

SECTION II

DESCRIPTION AND TABULATED DATA

	Paragraph
Description	4
Tabulated data	5

4. Description (figs. 3 to 5, incl.).—*a*. The medium tank is an armored, full-track-laying vehicle, powered by a nine-cylinder, air-

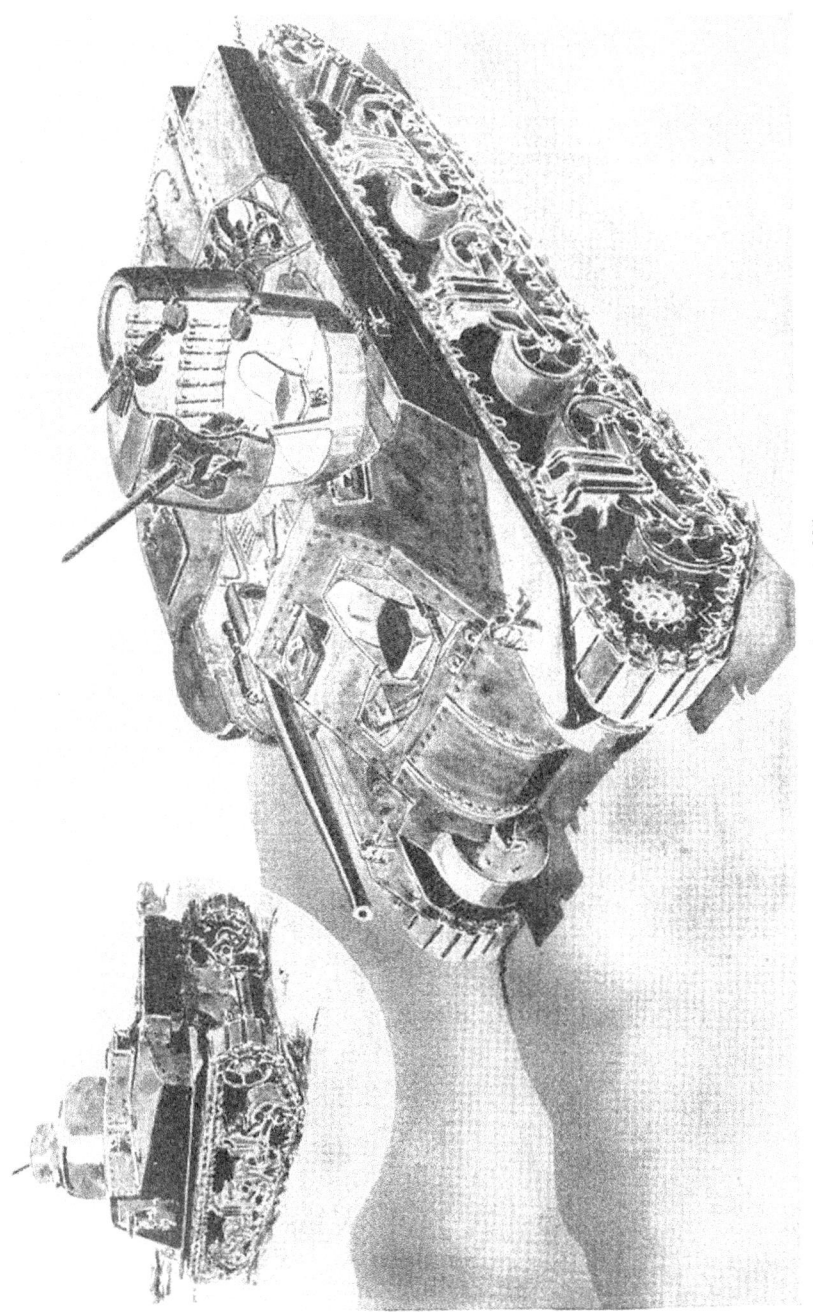

Figure 1.—Medium tank M3.

TM 9-750

ORDNANCE MAINTENANCE

Figure 2.—Right front view of tank.

Figure 3.—Left rear view of tank.

cooled radial aircraft type engine located in the rear of the hull. The medium tanks M3, M3A1, and M3A2 differ principally in the construction of the hull. The M3 has a riveted hull; the M3A1 has a cast and upper hull; the M3A2 tank hull is of welded construction. All information contained in this manual is applicable to all three of these models. All models may be powered by either a gasoline or Diesel engine. The operator steers the vehicle by means of two levers located in the front end of the hull. The vehicle has five forward speeds and one reverse. The tank is wired for radio installation and for an interphone system within the tank.

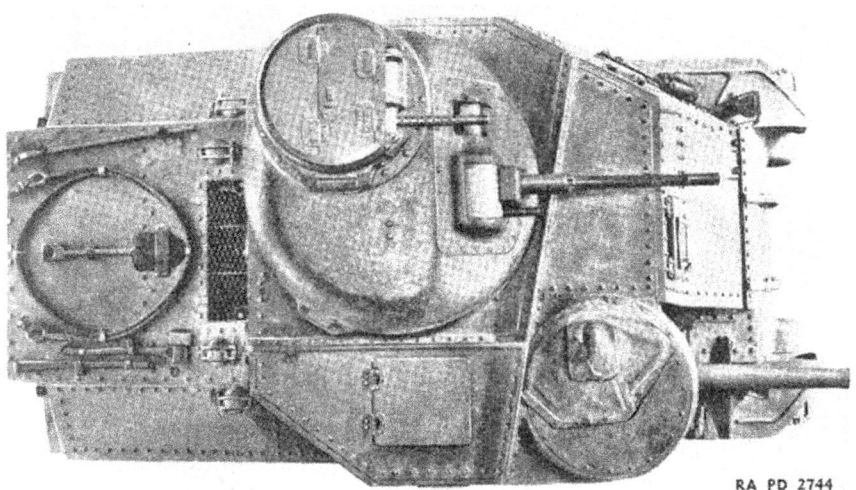

RA PD 2744

FIGURE 4.—Top view of tank.

b. The armor plate is considerably thicker than on previous medium tanks. The armor of the front upper section, cupola, and turret sides is 2 inches thick; the armor on the sides of the hull and the front lower section is 1½ inches thick.

c. The turret is 60 inches in diameter and can be rotated by means of a hydraulic system, or by hand. The cupola normally rotates with the turret but can be rotated independently by hand.

d. There is an auxiliary electrical generating system in the tank, consisting of a generating set powered by a single-cylinder gasoline engine. This unit charges batteries, heats the engine and fighting compartments, and supplies current to operate turret traversing mechanism and radio.

TM 9-750

ORDNANCE MAINTENANCE

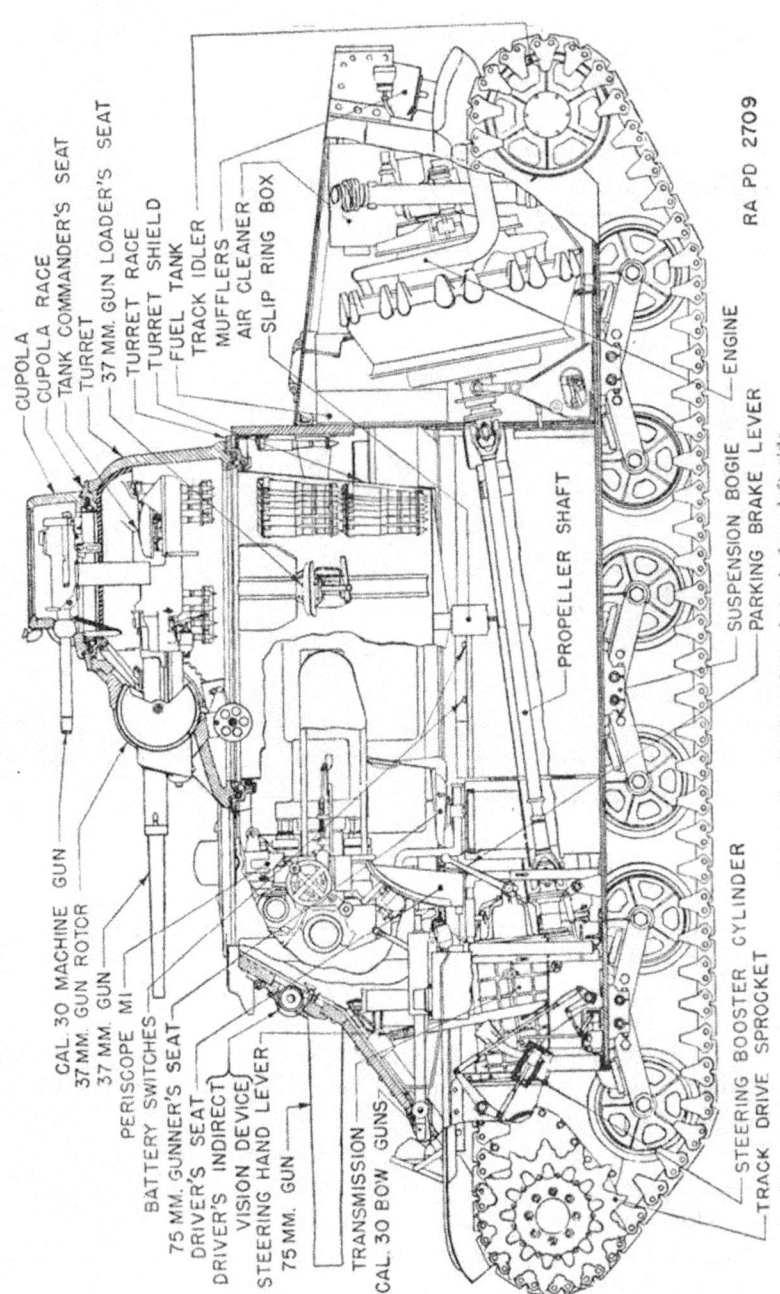

Figure 5.—Section through tank from left side.

MEDIUM TANKS M3, M3A1, AND M3A2

5. Tabulated data.—*a. General.*

Weight without armament, fuel, and crew_____pounds___ 55,600
Weight fully equipped_____tons, approximately__ 30
Ground pressure_____p. s. i__ 17.5
Over-all width_____feet___ 8^{11}/_{12}
Ground clearance_____inches__ 17
Tread (center to center of tracks) _____do____ 83½
Over-all height _____do_____do____ 10¼
Over-all length _____do____ 18½

b. Engine.—Wright Whirlwind R–975–EC 2.
Rated horsepower_____ 400
Number of cylinders_____ 9
Weight of engine with accessories_____pounds__ 1,370

c. Armament.

    1 75-mm gun.
    1 cal. .30 machine gun in cupola.
    1 37-mm gun and 1 cal. .30 machine gun mounted parallel to each other in turret.
    2 cal. .30 machine guns in front end.
    2 cal. .45 Thompson submachine guns carried on brackets within tank.

d. Ammunition carried.

    50 rounds 75-mm ammunition.
    178 rounds 37-mm ammunition.
    9,200 rounds cal. .30 ammunition.
    1,300 rounds cal. .45 ammunition in 50-round magazines.

e. Protected vision.—Protected vision is provided for the driver and other members of the crew by indirect vision devices called protectoscopes, installed in the doors and front of the tank.

f. Seats, body supports, and safety belts are provided for each of the seven members of crew.

g. Portions of the interior are padded with sponge rubber needed to protect the crew against bumps.

h. Communication.

(1) Radio_____ { SCR–245 sending and receiving:
                                                          _____miles__ Voice 15–25
                                                          _____do____ Code 30–45
(2) Intratank_____ Telephone

i. Armor thickness.

Front	2 in.
Rear	1½ in.
Sides	1½ in.
Top	½ in.
Bottom	½ in.

j. Turret.—Cast armor 2 inches thick, 360° traverse.

k. Fuel and oil.

Fuel capacity	gallons	185
Number of miles without refueling (8 hours at 18 mph)	miles	144
Octane rating of fuel		91 or higher
Oil consumption (approximate)		2 qt. per hour
Engine oil capacity	quarts	36
Lubricants		See figure 12

l. Performance.

Maximum sustained speed on hard road	mph	25
Expected cross-country speeds for various terrains	mph	4 to 20
Maximum allowable engine speed	rpm	2,100
Minimum engine idling speed	rpm	800
Maximum grade ascending ability	degrees	35
Maximum grade descending ability	do	35 to 45
Maximum width of ditch tank will cross	feet	7 5/12
Maximum vertical obstacle, such as a wall, that the tank will climb over	inches	24
Maximum fording depth (at slowest forward speed)	do	42

m. Crew.—Seven men.
n. Tracks.—Rubber block.

Track shoe width	inches	16
Track pitch	do	6
Ground contact	square inches	3,432
Blocks per track		79

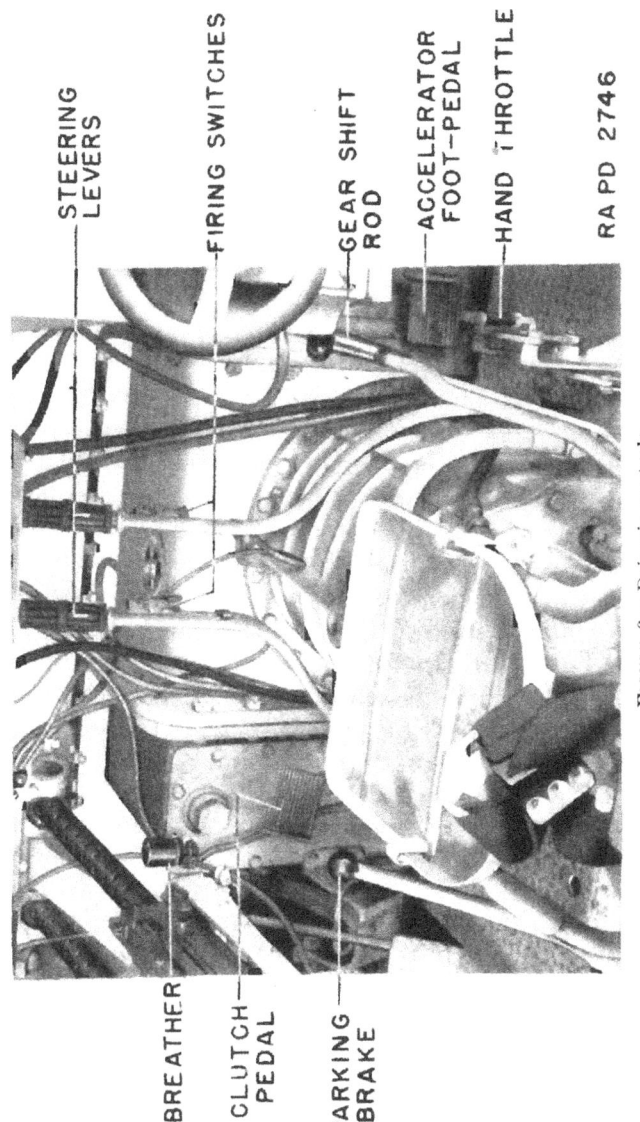

Figure 6.—Driver's controls.

Section III

OPERATION INSTRUCTIONS AND CONTROLS

	Paragraph
General information on controls	6
Inspection before operation	7
Starting instructions for gasoline engine	8
Starting instructions for Diesel engine	9
Engine test	10
Operating vehicle	11
Stopping engine	12
Cautions	13
Cold weather operation	14

6. General information on controls.—*a. Spark control.*—The spark control is entirely automatic and requires no attention by the operator of the vehicle.

b. Accelerator and hand throttle.—A foot accelerator pedal is located to the right of the transmission housing, convenient to the driver's right foot. In conjunction with the foot pedal, a hand-operated throttle is provided, which is mounted on the floor to the right of the operator and to the right of the gear shift lever.

c. Steering levers.—(1) Two steering levers are mounted on the floor of the vehicle, one on each side of the driver's seat. To steer the vehicle, pull the steering lever on the side toward which it is desired to turn. Pulling back either one of the levers slows down the track on that side, while the speed of the other track is increased. Thus the vehicle turns with power on both tracks at all times (fig. 6).

(2) The levers are provided with rubber grips. The lower ends are mounted in such way as to permit relocating the lever positions to suit the convenience of the driver. Both levers are equipped with switches for firing the machine guns in the front end of the tank.

d. Brakes.—(1) *Service brakes.*—Pulling back simultaneously on both steering levers slows down or stops the vehicle, depending on the effort applied.

(2) *Parking brake.*—The parking brake lever is located on the left side of the driver, back of the steering lever. *It is a transmission type brake, and should never be used for any purpose other than parking.*

e. Clutch.—The clutch pedal is located to the left of the transmission housing, convenient to the driver's left foot. To permit shifting of gears, the clutch is disengaged by depressing the clutch pedal. When the pedal is depressed, the engine will run idle.

f. Gear shifting.—Shifting of gears in the transmission for speed changes is accomplished by the gear shift hand lever, located on the transmission, to the right of the driver. The positions of the gear shift lever for the various speeds are shown in figure 7. The gear shift lever is equipped with a latch which prevents accidental shifting into first speed or reverse. The latch must be released by pressing down the button on top of the lever before shifting into first speed or reverse.

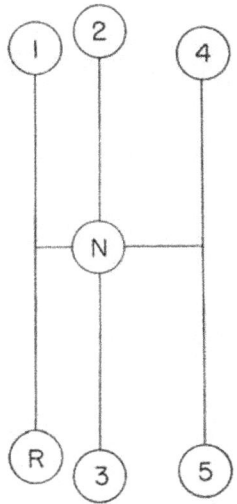

RA PD 2712

FIGURE 7.—Gear shift positions.

7. **Inspection before operation.**—*a. By driver.*—(1) Checks oil tank level, adds oil if necessary.

(2) Checks transmission oil level, adds oil if necessary.

(3) Checks for presence and condition of fire extinguishers.

(4) Checks for oil or fuel leaks on floor of fighting compartment.

(5) Sees that steering levers, clutch pedal, and gear shift lever operate freely and over the full range.

(6) Sees that battery switch is open and voltmeter reads zero.

(7) Sees that fuel shut-off valve is closed.

(8) Closes battery switch and sees that voltmeter reads 24 or more volts.

(9) Turns on lights and siren at order of tank commander.

b. By tank commander.—(1) Checks final drive oil levels.

(2) Checks gasoline level.

(3) Oils throw-out bearing.
(4) Walks around tank and inspects—
(a) For gasoline or oil leaks underneath tank.
(b) That outside accessories, pioneer tools, tow cable, shackles and shackle pins, etc., are present.
(c) General condition of sprockets, bogie wheels, spring guides, gudgeons, track supporting rollers, and idlers.
(d) Tracks for wear, tightness and tension, and connections for wear.
(e) Condition of wedges and wedge nuts.
(f) Condition and tightness of grousers, if used.
(5) Turns engine over by hand about two complete revolutions to clear cylinders of hydrostatic lock.
(6) Causes driver to operate all lights and the siren.
(7) Sees that ammunition, flags, field equipment, and rations, when carried, are properly loaded.
(8) Checks elevation and traversing of vehicle weapons.
(9) Tests turret and cupola to see if it turns freely, by hand, and that locking and traversing mechanism function.
(10) Starts Homelite generator and then operates turret by power, checking for ease of operation.
(11) Has radio and antenna checked for operation.

8. Starting instructions for gasoline engine.—After the pre-starting inspection has been completed the engine is ready to start. This is done as follows:

a. Give the engine several strokes of the priming pump located on the instrument panel (fig. 8). Excessive priming should be avoided, since it has a tendency to wash lubricant off the cylinder walls and thus cause excessive wear and damage. Do not prime through exhaust ports or spark plug holes with raw gasoline. Do not prime a hot motor.

b. Move the hand throttle lever almost to the "closed" position.

c. Operate the starter. Turn the ignition switch to the "both" position and press the booster coil switch located on the instrument panel (fig. 8).

d. As soon as the engine fires, open the throttle slowly until the tachometer registers 800 rpm. If the throttle is opened rapidly when the engine starts to fire, the engine may stop.

e. After the engine has started—
(1) Check all instruments to see that they are functioning and showing the proper readings.
(2) Check for loose parts.

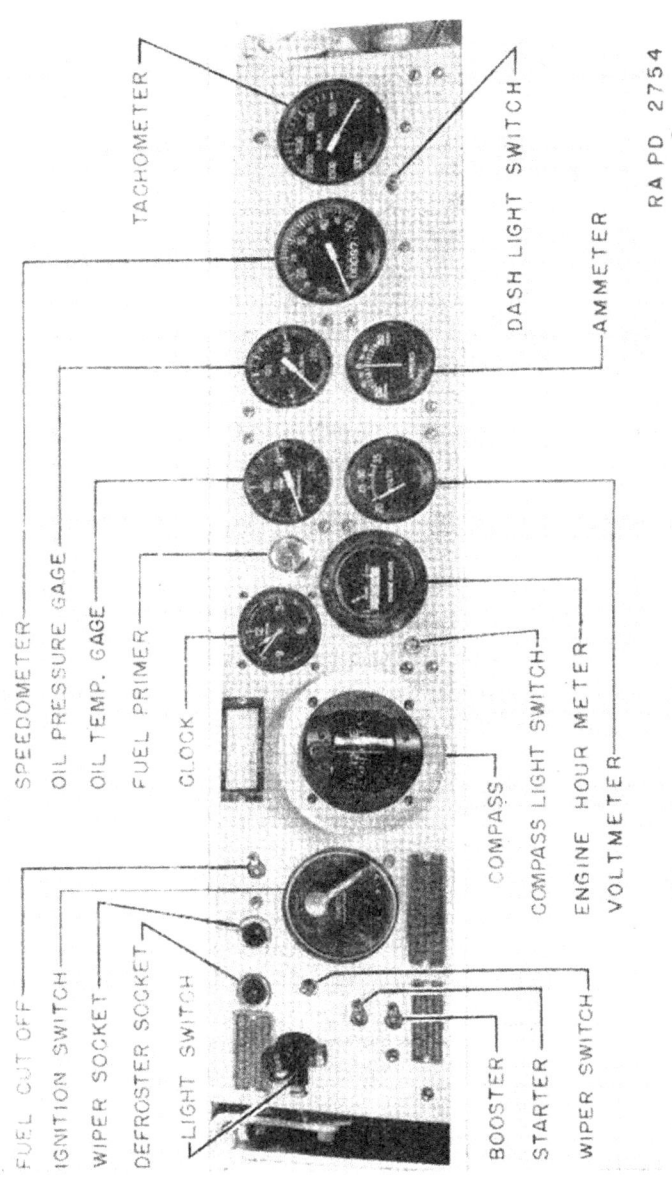

Figure 8.—Instrument panel.

(3) Check for unusual noises in power train and engine.

(4) Check magneto switch on 3 positions.

9. Starting instructions for Diesel engine.—After prestarting inspection, the Diesel engine is started as follows:

a. Actuate the decompression lever to put the engine on decompression.

b. Turn the engine one complete revolution by hand.

c. Put the engine back on compression.

d. Insert the proper cartridge in the cartridge starter and fire.

10. Engine test.—*a.* As soon as the engine has started, the oil pressure gage should be watched. If the gage does not indicate oil pressure within one-half minute, the engine should be shut down and an investigation made. In cold weather, when the oil has not been preheated, keep the engine throttled to 800 rpm until the oil temperature gage reads 80°.

b. Warm the engine up at a speed of 800 rpm until the oil pressure is between 50 and 80 pounds, which is normal for this speed. This warm-up should be continued for at least 5 minutes, after which the speed may be increased to 1,000 rpm. Should oil pressure fall off at an increase in speed, the engine should be throttled back for further warm-up before speed is again increased. A thorough warming up is recommended.

c. When the oil temperature gage reads 80°, open the hand throttle gradually until the tachometer reads 2,000 rpm and then close the throttle gradually until the engine speed drops to 800 rpm. Check for smooth, easy acceleration of engine. Check the oil pressure and the oil temperature and note the loss of speed when the engine is switched to one magneto at a time. The speed should not drop more than 75 rpm below that attained when operating on both magnetos.

11. Operating vehicle.—*a. General.*—The driver should be thoroughly familiar with the function and operation of all the controls and instruments before attempting to drive the vehicle. Review of paragraph 6 should be made in case of doubt.

b. Operation.—With the driver in the driver's seat, the engine at idling speed, and all instruments showing normal readings, the driver is then ready to drive the vehicle.

(1) Release the parking brake (located on floor to the left of driver's seat).

(2) Disengage the clutch by pressing the clutch pedal down to the floor and holding it down.

(3) Depressing the latch, move the gear shift lever into second gear, as shown in figure 7.

(4) After speeding up the engine, gradually release the clutch pedal. Except when under fire, attempting to move the vehicle in or out of close quarters without the aid of personnel outside of the vehicle serving as a guide is very dangerous.

(5) The correct gear for running is that which enables the vehicle to proceed at the desired speed without causing the engine to labor. Do not ride the clutch. The driver's left foot must be completely removed from the clutch pedal while driving, to avoid unnecessary wear on the clutch.

(6) To place the vehicle in reverse gear a complete stop must be made and the throttle closed until the tachometer reads 400 rpm (minimum idling speed). Shift briefly into third gear to stop the propeller shaft. Then depressing the latch move the gear shift lever to the reverse position (fig. 7). Backing the vehicle should never be attempted unless an observer is stationed in front to guide the driver.

(7) To steer, pull back the right-hand steering lever to make a right turn or the left-hand lever for a left turn. This action keeps one of the tracks from turning as fast as the other track and more power is needed. As the driver anticipates making a turn he must be ready to apply the foot throttle to a greater extent depending on the sharpness of the turn.

(8) To stop the vehicle, release the throttle and pull back on both steering levers at the same time, depressing the clutch when the vehicle has slowed down to approximately 2 to 5 miles per hour, depending upon which gear is being employed before stopping. It is desirable to shift down as low as possible, preferably to second gear, and use engine drag to slow the vehicle to facilitate stopping.

(9) The parking brake is located on the floor to the left of the driver's seat. This brake should be used only for parking and never while the vehicle is in motion. The operator should always make sure that this brake is released before putting the vehicle in motion.

(10) The tachometer, the oil temperature gage, and the oil pressure gage give the most satisfactory indications of the engine's performance. Should the indications of any of these instruments appear to be irregular, the engine should be stopped and the cause investigated. The oil temperature should not exceed 190°.

12. **Stopping engine.**—*a. Gasoline engine.*—To stop the engine close the throttle until the engine is idling at 800 rpm and run the engine at this speed for 4 or 5 minutes. Then throw the fuel cut-off

switch on the instrument panel to the "off" position. When the engine has stopped, move the ignition switch to the "off" position, and shut off the main fuel supply valves.

b. Diesel engine.—To stop the Diesel engine idle at 800 rpm for 4 or 5 minutes. Put the hand throttle in the extreme closed position.

Caution.—Do not shut off the fuel line valves. If these valves are shut off, it is possible for air to get into the injecting system, resulting in air locks.

13. **Cautions.**—*a.* After initial warming, the tank engine must be operated at appreciable engine speeds (1,800–2,100 rpm), except for brief periods at 1,500 to 1,600 rpm. Continuous operation of this type engine at idling speed will shorten the useful life of the engine considerably by causing increased wear and overheating. On no occasion except shifting gears, including military ceremonies, will idling of the engine at less than 800 rpm be permitted. While the damage to the engine is not immediately apparent, the total life of the engine will be greatly reduced.

b. The engine should be idled at about 800 rpm until the engine oil temperature gage indicates a temperature rise of about 10° F. before operating the vehicle at full speed and under load. This is necessary to prevent damage to the engine working parts. Avoid rapid movement of the accelerator, since this causes a spray of gasoline to be injected into the cylinders. This gasoline washes oil from the cylinder walls and causes excessive wear.

c. Do not attempt to start the engine by "towing" or "coasting" the vehicle. To do so may cause serious damage to the engine and transmission. To start the engine, use the electric starter or hand starter.

d. Care must be taken not to place any object in a position where it will block the flow of air from the cylinders. Blocking off any cylinder will cause overheating and preignition in the cylinder so affected. Care must also be taken not to block engine oil cooler.

14. **Cold weather operation.**—In order that the tank engine may be started and operated with a minimum of difficulty in very cold weather it will be necessary that the following precautions be taken prior to starting.

a. Before stopping the engine in cold weather, pull out the oil dilution valve operating button, which is located under the engine oil tank, and hold the valve open the appropriate number of seconds, depending upon the anticipated temperature (see chart in par. 17). The older models have manually operated valves; later models have a solenoid-operated valve which is controlled by a button on the instrument panel. In case the valve is manually operated, the engine compartment must be opened and the valve checked to make sure it is closed. Following a shut-down on diluted oil, the engine must be operated at least 20 minutes when next started to insure the evaporation of the gasoline in the oil. The oil dilution line must be disconnected at all times except during weather when dilution is absolutely necessary and then the oil dilution system must not be used except when cold starting conditions are anticipated.

b. If closed buildings or shelters of any kind are available, the tank should be placed in these shelters and some method of heating provided. Take due precautions against fire hazards from leaking fuel or oil. If such shelters are not available, tanks which must be kept available for immediate use may be kept warm considerably above prevailing temperatures by covering with a tarpaulin.

c. A short time before starting the engine, it should be heated by means of the auxiliary generator set according to the directions given in section XII, chapter 2.

d. Under extremely cold conditions it may be necessary, in order to avoid malfunctions, to exercise special precautions in lubricating machine guns and other armament. TM 9–850 gives information covering the proper treatment of such equipment during extremely cold weather.

Section IV

LUBRICATION

	Paragraph
General	15
Methods	16
Oil dilution valve	17
Lubrication instructions for medium tanks M3 and M3A1	18

15. General.—Figure 12 shows the various points to be lubricated, the periods of lubrication, and lubricants to be used. In addition to the items on the chart, other moving parts such as door and shield hinges, pistol port covers, door latches, and gun mounting pins must be lubricated at frequent intervals. Oilholes and lubrication fittings are painted red for easy identification.

16. Methods.—*a.* In the engine oil system (fig. 9), oil is drawn from the bottom of the oil tank by the engine oil pump and forced into the engine oil circuit. The excess oil in the engine oil circuit is returned directly to the oil tank by the oil pump. The oil, after lubricating the engine parts, accumulates at the bottom of the crankcase, and this oil is pumped by the oil scavenging pump through the oil filter to the oil cooler and then to the oil tank. The oil tank and oil cooler are located at the bulkhead in the fighting compartment.

TM 9-750
ORDNANCE MAINTENANCE

b. The handle on the disk type oil filter (fig. 10) should be turned *one complete* revolution (in either direction) each day. If examination of the oil shows any appreciable amount of dirt or discoloration, the filter should be removed and exchanged for a new one.

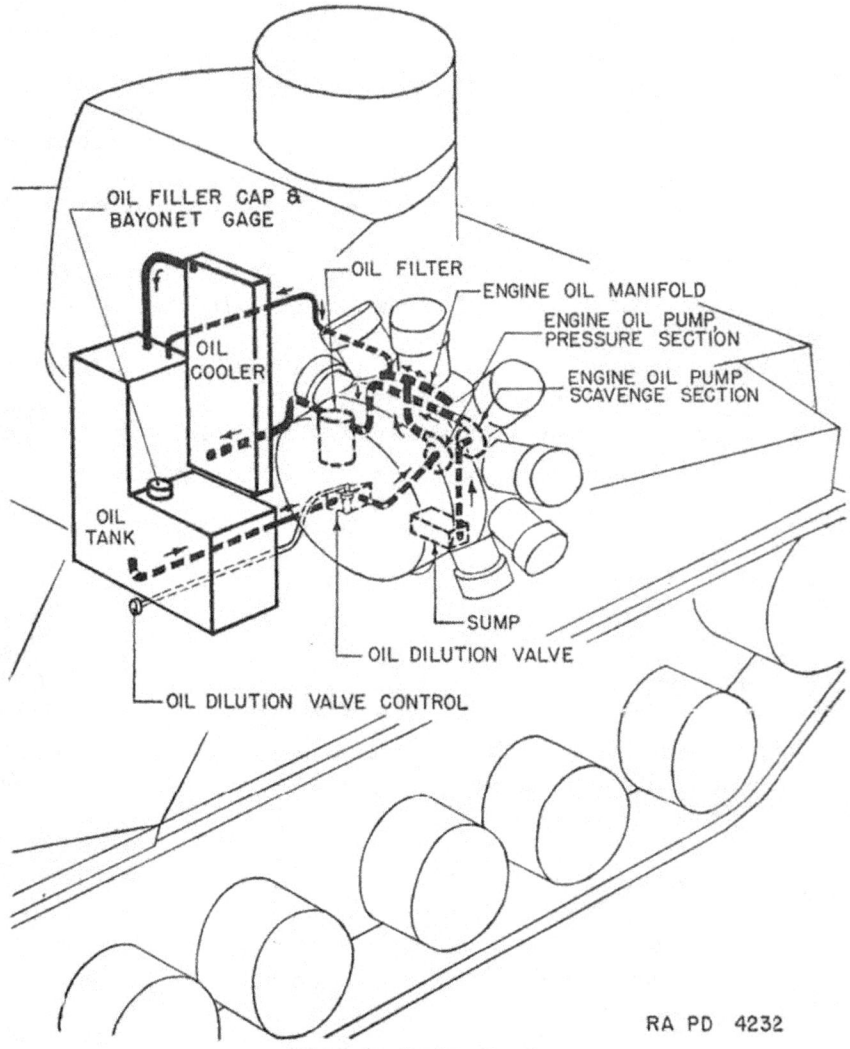

FIGURE 9.—Engine oil system.

c. There is one breather connection, from the nose of the engine to the left air intake pipe, to provide necessary venting of the engine.

d. The oil tank (fig. 11) is filled through the filler hole on the tank. The cap for this hole is fitted with a bayonet gage which indicates

MEDIUM TANKS M3, M3A1, AND M3A2

TM 9-750
16-17

the oil level. To drain the system, remove the cover plate located below the hull floor plate directly under the oil tank. This permits removal of the drain plug.

17. **Oil dilution valve.**—*a.* This valve is used to permit easier starting of the engine in cold weather by mixing gasoline with the engine oil. Before stopping the engine in cold weather, pull out the

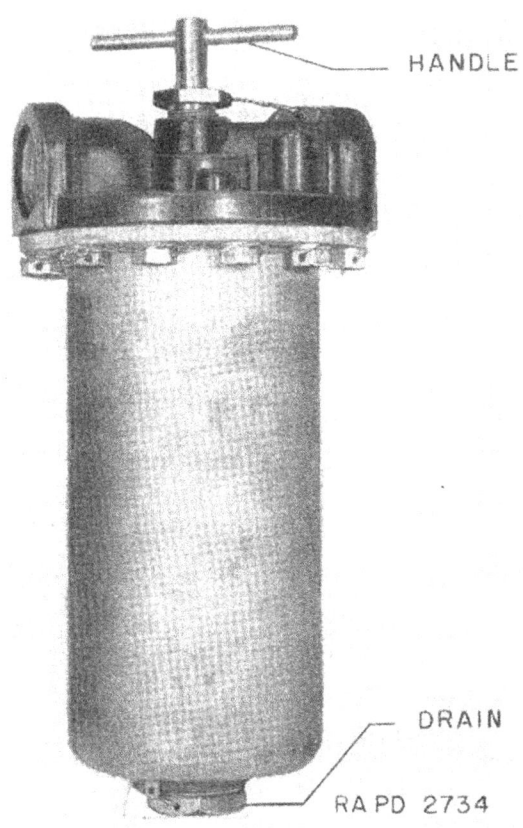

FIGURE 10.—Disk type oil filter.

valve operating button, which is located under the engine oil tank (fig. 11). The button opens the gasoline valve and allows gasoline to mix with the oil. The period of time to hold the valve open will depend on the temperature. To determine the period of time to hold the valve open, estimate the lowest temperature anticipated at the next engine starting. Find this temperature in the table given

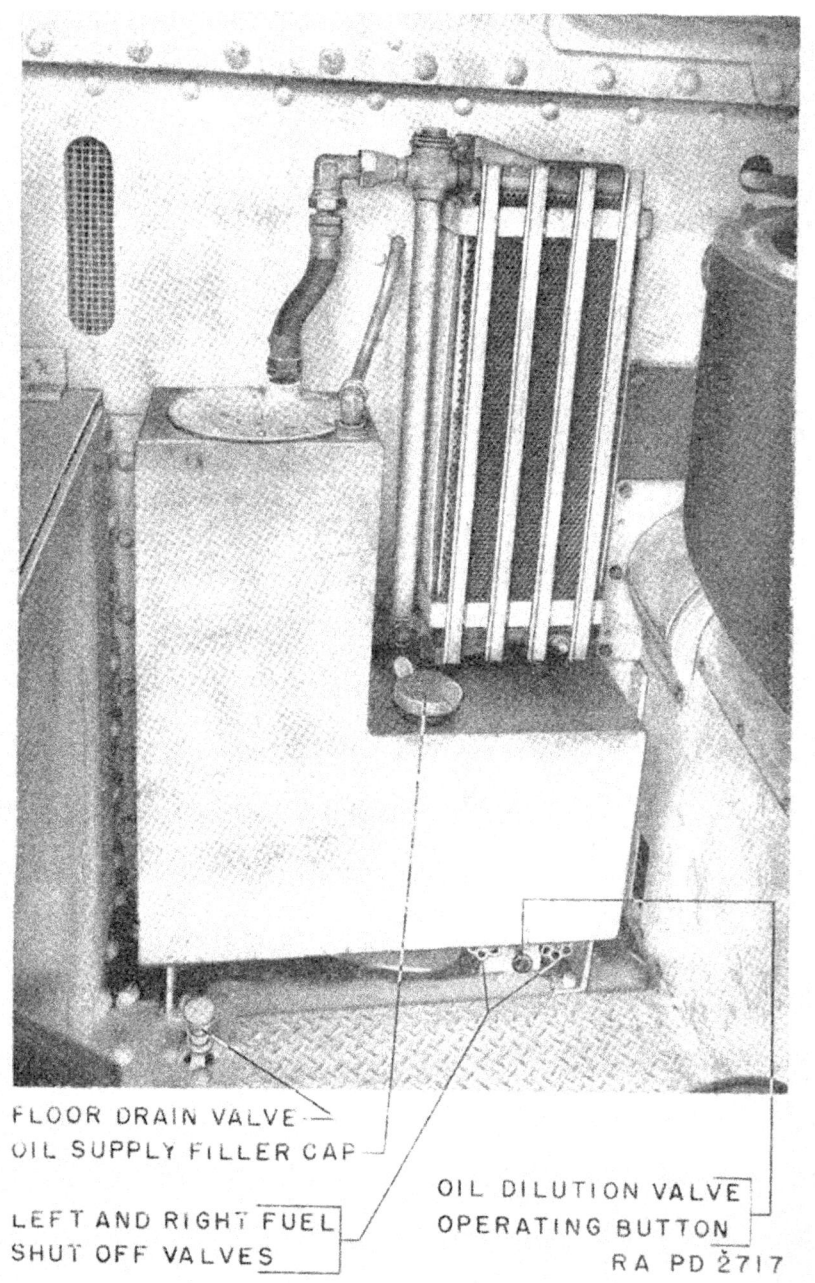

FIGURE 11. — Oil tank and cooler.

MEDIUM TANKS M3, M3A1, AND M3A2   17-18

below and hold the oil dilution valve open the length of time indicated opposite this temperature in the chart.

Temperature (° F.)	Dilution time (seconds)
+30	8
+20	30
+10	60
0	90
—10	130
—20	175
—30	250

During the time the valve is held open, idle engine at 1,000 rpm. Shut the engine off immediately after the dilution valve is closed. Time in seconds may be estimated by counting one thousand one, one thousand two, and so on. Each expression "one thousand one", etc., represents approximately 1 second.

b. Important points to remember are—

(1) Engine *must* be operated for at least 30 minutes when started following a shut-down on diluted oil.

(2) Engine *must* be shut down *immediately* after dilution.

(3) Dilution time greater than 250 seconds must *not* be used.

(4) Dilution button or valve must not be used except during dilution period.

18. **Lubrication instructions for medium tanks M3 and M3A1.**—See figure 12.

Table of capacities and recommendations

	Capacity (quarts)	Above 90°	Lowest expected atmospheric temperature				
			+32°	+10°	—10°	—30°	Below —30°
Engine oil tank:							
Gasoline	36						
Diesel	50	SAE 50	SAE 50	SAE 50	SAE 50	For operation in these temperature ranges, refer to OFSB 6-G-3.	
Transmission	26						
Differential and final drive	120						

21

TM 9-750

ORDNANCE MAINTENANCE

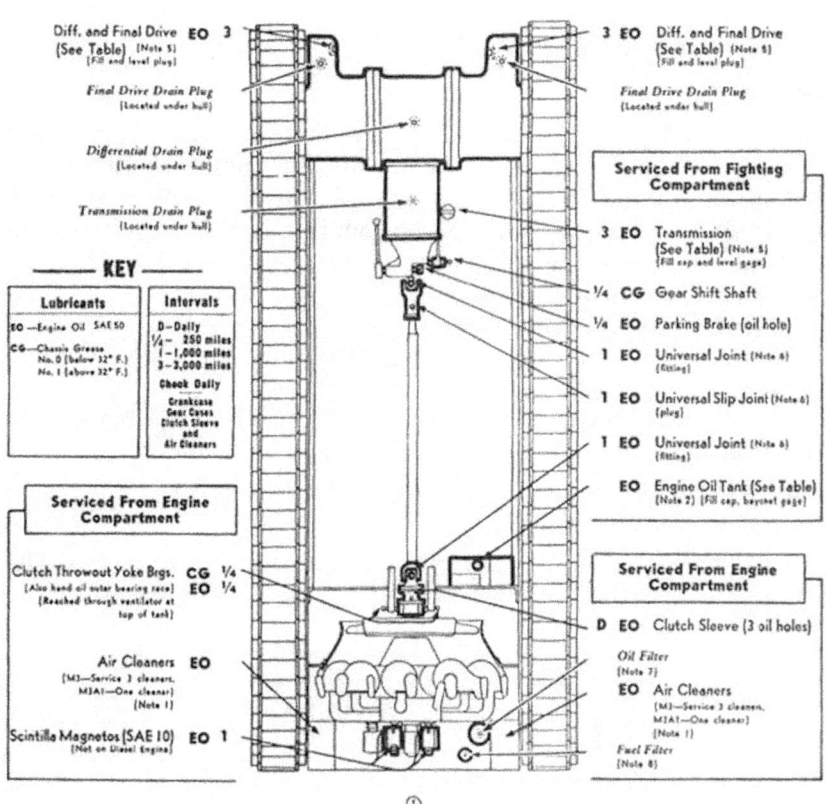

Figure 12.—Lubrication instructions for medium tanks M3 and M3A1. (Ordnance Department starting serial No. 1 located on name plate inside fighting compartment; SNL G-104.)

MEDIUM TANKS M3, M3A1, AND M3A2   TM 9-750
                                       18

NOTES

1. *Air cleaners.*—Drain, clean, and refill with engine oil SAE 30 above 32° F., SAE 10 below 32° F., daily when operating on dirt roads or across country. Service every 250 miles when operating on paved roads or during wet weather. Depending upon operating conditions, remove air cleaner and wash all parts every 100 to 500 miles. *Caution:* Keep all air pipe connections tight.

2. *Engine oil tank.*—Check oil level daily. Drain at intervals shown below, only when engine is hot. Clean tank and oil filling tube strainer every 1,000 miles. Refill oil tank to "full" mark on bayonet gage, located under fill cap. *Caution:* Do not remove strainer when filling tank.

   a. Gasoline engine.—Drain and refill every 250 miles or 25 hours when used for cross-country or dirt road operation. Drain and refill every 1,000 miles or 100 hours for paved road or wet weather operation.

   b. Diesel engine.—Drain and refill every 250 miles or 25 hours of operation under all operating conditions.

3. *Intervals.*—Intervals indicated are for normal service. For extreme conditions of speed, heat, water, mud, snow, dust, etc., change engine oil and lubricate more frequently.

4. *Fittings.*—Clean before applying lubricant. Lubricate until new grease extrudes from the bearing.

5. *Gear cases.*—Check level daily; add lubricant if necessary. Check with tank on level ground. Drain, flush, and refill at the end of first 250 miles; thereafter as indicated at points on guide. Clean transmission and differential filler strainer every 3,000 miles. *Caution:* Do not remove strainer when filling.

6. *Universal joint and slip joint.*—Remove tunnel shield sections over universal joints and slip joint at ends of tunnel shield. To lubricate slip joint, remove plug and insert fitting. Apply EO to universal joint until it overflows at relief valve and to slip joint until EO extrudes from end of spline. *Caution:* After lubricating remove fitting and replace plug.

7. *Oil filter.*—Turn handle on top of filter one full turn daily. Drain every 250 miles. Some M3A1 models have automatic type filter.

8. *Fuel filter.*—*a. Gasoline.*—Turn handle on top of filter one full turn daily. Drain every 250 miles.

   b. Diesel.—Open vent cocks, drain through drain plug, and remove filter element. Wash element and case, replace element, and prime fuel system. Replace second stage American Bosch filters on M3A1 models every 3,000 miles.

9. *Oilcan points.*—Lubricate door and shield hinges, peephole protectors, pistol port covers, door latches, control rod pins and lever bushings, etc., with engine oil SAE 30 every 250 miles.

10. *Points to be lubricated at time of engine removal for periodic inspection.*—Clutch pilot bearing and clutch hub bearing.

11. *Points to be lubricated by ordnance maintenance personnel at time of general engine overhaul.*—Clutch pilot bearing, clutch hub bearing, generator, magneto (except oilers), and starter (see OFSB 6-G-104).

TM 9-750
ORDNANCE MAINTENANCE

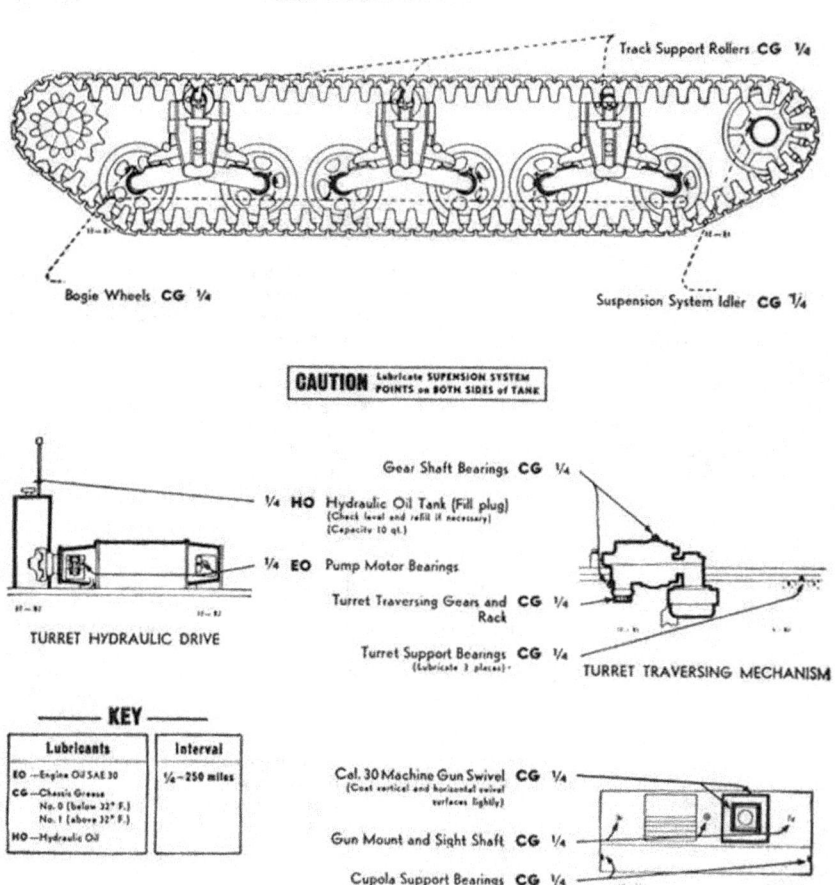

FIGURE 12.—Lubrication instructions for medium tanks M3 and M3A1—Continued.

NOTES

1. *Intervals.*—Intervals indicated are for normal service. For extreme conditions of speed, heat, water, mud, snow, rough roads, dust, etc., lubricate more frequently.

2. *Fittings.*—Clean before applying lubricant. Lubricate bogie wheels, idler, and track support rollers until lubricant overflows relief valve. Lubricate other fittings until new grease extrudes from the bearing. *Caution:* Lubricate suspension points after washing vehicle.

3. *Portable generator.*—Two-cycle, air-cooled engine, mounted on turret floor rear, is lubricated by engine oil mixed with the fuel. Mix thoroughly ⅜ pint EO with each gallon of gasoline before pouring into tank. *Caution:* Do not pour gasoline and oil separately into tank. Keep fuel strainer clean. Every 200 hours lubricate magneto cam follower by oiling felt with one or two drops EO.

4. *Oilcan points.*—Lubricate peephole protector slides, door hinges and latches, shield hinges, control pins and clevises, etc., with EO sparingly every 250 miles.

5. *Points requiring no lubrication.*—Turret traversing mechanism gears, bogie wheel suspension linkage and slides, and final drive sprocket bearings.

Section V
INSPECTIONS

	Paragraph
Purpose	19
Inspection before operation	20
Inspection during operation	21
Inspection at halt	22
Inspection after operation	23
Periodic inspection	24

19. Purpose.—*a.* To insure mechanical efficiency, it is necessary that tanks be systematically inspected at intervals in order that defects may be prevented or discovered and corrected before they result in serious damage.

b. Cracks that develop in castings or other metal parts may often be detected on the completion of a run by an examination for dust and oil deposits around the cracks.

c. Suggestions toward changes in design prompted by chronic failure or malfunction of a unit or group of units; pertinent changes in inspection or maintenance methods; and changes involving safety, efficiency, economy, and comfort should be forwarded to the office of the Chief of Ordnance, through technical channels, at the time they develop. Such action is encouraged, in order that other organizations may profit thereby.

20. Inspection before operation.—*a. By driver.*—(1) Checks oil tank level; adds oil if necessary.

(2) Checks transmission oil level; adds oil if necessary.

(3) Checks for presence and condition of fire extinguishers.

(4) Checks for oil or fuel leaks on floor of fighting compartment.

(5) Sees that steering levers, clutch pedal, and gear shift lever operate freely and over the full range.

(6) Sees that battery switch is open and voltmeter reads zero.

(7) Sees that fuel shut-off valve is closed.

(8) Closes battery switch and sees that voltmeter reads 24 or more volts.

(9) Turns on lights and siren at order of tank commander.

(10) Inspects to see that hydraulic brake booster pump is clear of engine exhaust pipe.

(11) Removes inspection plate and examines "tee" connection of brake booster system, located forward of engine, for leaks.

(12) Examines all hydraulic lines of brake booster system for chafing and loose anchor clips.

(13) Checks level of sump tank of brake booster system.

b. By tank commander.—(1) Checks final drive oil levels.

(2) Checks gasoline level.

(3) Oils throw-out bearing.

(4) Walks around tank and inspects—

(*a*) For gasoline or oil leaks underneath tank.

(*b*) That outside accessories, pioneer tools, tow cable, shackles and shackle pins, etc., are present.

(*c*) General condition of sprockets, bogie wheels, spring guides, gudgeons, track supporting rollers, and idlers.

(*d*) Tracks for wear, tightness, and tension; end connections for wear.

(*e*) Condition of wedges and wedge nuts.

(*f*) Condition and tightness of grousers, if used.

(5) Turns engine over by hand about two complete revolutions to clear cylinders of hydrostatic lock.

(6) Causes driver to operate all lights and the siren.

(7) Sees that ammunition, flags, field equipment and rations, when carried, are properly loaded.

(8) Checks elevation and traversing of vehicle weapons.

(9) Tests turret and cupola to see that it turns freely, by hand, and that locking and traversing mechanism function.

(10) Starts Homelite generator and then operates turret by power, checking for ease of operation.

(11) Has radio and antenna checked for operation.

21. Inspection during operation.—*a.* During operation the driver should be alert to detect abnormal functioning of the engine. He should be trained to detect unusual engine sounds or noises. He should glance frequently at the instrument panel gages to see if the engine is functioning properly. An unsteady oil gage pointer indicates low oil pressure, provided that engine speed is fairly constant. The steering mechanism must be checked for clearance before engagement, intensity of pull required for braking, etc.

b. Only under exceptional circumstances should a tank be operated after indications of trouble have been observed. When in doubt, the engine should be stopped and assistance obtained. Inspection during operation applies to the entire vehicle and should be emphasized throughout the driving instruction period.

22. Inspection at halt.—*a.* At each halt the operator should make a careful inspection of the tank to determine its general mechanical condition. Minor defects detected during the march together with defects discovered at the halt should be corrected before resuming the march. If the defects cannot be corrected during the halt, proper disposition of the vehicle should be made so that unnecessary delay may be avoided and major failure prevented.

  b. A suitable general routine is as follows:

  (1) Allow the engine to run a short time at idling speed (800 rpm). Listen for unusual noises.

  (2) Note any unusual gear train noises.

  (3) Check for transmission and differential oil leaks inside fighting compartment.

  (4) Note steering adjustment.

  (5) If barrels on machine guns have worked loose, they should be adjusted.

  (6) Through the opening in the bulkhead examine the clutch throwout bearings and oil their outer race.

  (7) Walk around the vehicle, looking carefully for fuel or oil leaks.

  (8) Remove dirt from under the track support rollers and from between bogie wheel arms.

  (9) Examine tracks for adjustment and for worn, loose, broken, or missing parts.

  (10) Inspect hull and fittings for missing, worn, or loose parts.

  (11) Feel steering brake housings and gear case for evidence of overheating.

  (12) Open doors to engine compartment and inspect for loose parts, loose connections, and leaks.

  (13) Inspect the lights, if traveling at night with lights.

  (14) Check the amount of fuel in the tank.

  (15) Wipe all windshields and vision devices. Do not use an oily or dirty rag.

23. Inspection after operation.—At the conclusion of each day's operation the tank commander should cause an inspection to be made similar to that made at halts but more thorough and detailed, as indicated by the Tank Commander's Report. The inspection should be followed by preventive maintenance. If defects cannot be corrected, they should be reported promptly to the chief of section or other designated individual. The following points should be covered:

  a. Examine the tracks and bogies.

  b. Check track tension.

  c. Inspect idler and roller tires.

  d. Examine the drive sprockets for worn or broken teeth.

  e. Examine the rubber track shoe units for unserviceable units.

  f. Check transmission oil level after tank has stood overnight.

  g. Check differential oil level after tank has stood overnight.

h. Check and clean air cleaners during extremely dusty operations.

i. Inspect light, siren, and windshield wipers. Check for loss or damage of exhaust mufflers and accessories.

j. Inspect the sighting and vision devices for breakage.

k. Inspect guns and mounts for defective performance.

l. Inspect guns, sighting equipment, and accessories to determine whether covers are properly installed.

m. Inspect ammunition and sighting compartments for cleanliness and orderly arrangement.

n. Replenish oil, fuel, and ammunition.

o. For continuous operation in hot weather, battery water must be replenished about twice a week.

p. Check operation of turret.

q. Test auxiliary generating unit.

r. Check brake booster pressure gage, which should read above 1,450 psi.

s. Check oil level of booster system sump tank.

t. Check entire brake booster system for visible leaks.

u. Remove empty shells, twigs, etc., from moving parts at brake booster control valve.

v. Check tightness of steering brake bands and action of hydraulic boosting by operating brake levers.

24. Periodic inspection.—The following periodic inspections are prescribed:

a. Daily check.

By driver	By tank commander
Transmission oil level.	Amount of fuel.
Gasoline shut-off valves.	Engine oil level.
Battery switch.	Turn over engine by hand.
Fire extinguishers.	Pioneer tools.
Engine tools.	Tow cable.
Steering levers.	Shackle and shackle pins.
Clutch pedal and free travel.	Guides and gudgeons.
Control linkage.	Bogies.
Gear shift lever.	Springs.
Interlock retaining screw and plug.	Idlers.
Primer.	Track supporting rollers.
Starter.	Sprockets.
Accelerator.	Tracks.
Instrument readings at 1,000 rpm.	End connectors.
Ammeter.	Wedges and wedge nuts.
	Grousers.

MEDIUM TANKS M3, M3A1, AND M3A2

By driver	By tank commander
Voltmeter.	Equipment.
Oil pressure gage.	Turret, ports, etc.
Temperature gage (engine warm).	Lights.
	Siren.
Idles at 800 rpm.	Shielding.
Magnetos and plugs.	Ammunition.
Clutch release bearing oiled.	Weapons.
Transmission oil petcock.	Gun mounts.
Engine noises.	Oil leaks.
Transmission noises.	Fuel leaks.
	Air horn connection.
	Engine noises.
	Is tank clean?
	Turn Cuno filter one turn.

b. 50-hour check.—(1) Remove, clean, and test spark plugs.

(2) Remove magneto breaker point housings and check points.

(3) Remove carburetor strainer, bypass strainer, and oil pump screen.

(4) Check carburetor mounting bolts, air intake pipes for loose hose clamps, leaks, or cracked carburetor elbow.

(5) Check engine for any loose parts, nuts, bolts, lines, connections, etc.

(6) Remove any dirt accumulated under two lower cylinder heads.

(7) Check carburetor throttle controls for missing cotter keys and for operation.

(8) Check governor for operation and rpm (2,100).

(9) Check oil and gas line connections, shut-off valves, etc., for leaks throughout.

(10) Remove air cleaners *completely* and clean.

(11) Check clutch operation and clutch release bearings for proper clearance.

(12) Repeat daily check steps and make our driver's report.

c. 100-hour check.—(1) Check all items on 50-hour and daily checks. Thoroughly check lubrication of the vehicle (follow lubrication chart).

(2) Remove engine from tank. (Following paragraph 59, *break all connections where specified—follow all steps in order.*)

(3) Remove, disassemble, and clean Cuno oil filter.

(4) Remove and clean oil sump strainer.

(5) Flush oil cooler and lines.

(6) Drain oil tank and remove hopper from tank. Inspect for sludge and dirt. Clean.

(7) Disassemble clutch, clean it, and check for cracked or worn plates.

(8) Remove radio operator's seat, tunnel inspection plate, and tunnel cover (rear). Break slip joint and front companion flange bolts and remove propeller shaft.

(9) Check joint in accelerator linkage. Clean out tunnel. Check lines and linkage in tunnel.

(10) Clean fighting and engine compartments and trace cause of any excess oil.

(11) Remove rocker box covers and check all valve clearances.

Section VI

GENERAL CARE AND PRESERVATION

	Paragraph
Records	25
Cleaning	26
Painting	27
Preparing for painting	28
Painting metal surfaces	29
Paint as camouflage	30
Removing paint	31
Painting lubricating devices	32

25. Records.—*a. Use.*—An accurate record must be kept of each motor vehicle issued by the Ordnance Department. For this purpose the Ordnance Motor Book (O. O. Form No. 7255), generally called "Log Book," is issued with each vehicle and must accompany it at all times. This book furnishes a complete record of the vehicle, from which valuable information concerning operation and maintenance costs, etc., are obtained, and organization commanders must insist that correct entries are made. This book will habitually be kept in a canvas cover to prevent its being injured or soiled.

b. The page bearing a record of assignment must be destroyed prior to entering the combat zone. All other references which may be posted regarding the identity of the organization must also be deleted.

26. Cleaning.—*a.* Grit, dirt, and mud are the sources of greatest wear to a vehicle. If deposits of dirt and grit are allowed to accumulate, particles will soon find their way into bearing surfaces, causing unnecessary wear, and if the condition is not remedied will soon cause serious difficulty. When removing engine parts or any other units, in making repairs and replacements, or if in the course of inspection working joints or bearing surfaces are to be exposed, all dirt and grit that might find its way to the exposed surfaces must first be carefully removed. The tools must be clean, and care must always be taken to eliminate the possibility of brushing dirt or grit into the opening with the sleeve or other part of the clothing. To cut oil-soaked dirt and grit, hardened grit, or road oil use dry-cleaning solvent applied with rags (not waste) or a brush. The vehicle is so designed that the possibility of interfering with its proper operation by careless application of cleaning water is very small. However, care should be taken to keep water from the power unit, as it might interfere with proper ignition and carburetion.

b. Oil holes which have become clogged should be opened with a piece of wire. Wood should never be used for this purpose, as splinters are likely to break off and permanently clog the passages. Particular care should be taken to clean and decontaminate vehicles that have been caught in a gas attack. See section VII for details of this operation.

27. Painting.—*a.* Ordnance matériel is painted before issue to the using arms and one maintenance coat per year will ordinarily be ample for protection. With but few exceptions this matériel will be painted with enamel, synthetic, olive-drab, lusterless. The enamel may be applied over old coats of long oil enamel and oil paint previously issued by the Ordnance Department if the old coat is in satisfactory condition for repainting.

b. Paints and enamels are usually issued ready for use and are applied by brush or spray. They may be brushed on satisfactorily when used unthinned in the original package consistency or when thinned no more than 5 percent by volume with thinner. The enamel will spray satisfactorily when thinned with 15 percent by volume of thinner. (Linseed oil must not be used as a thinner, since it will impart a luster not desired in this enamel.) If sprayed, it dries hard enough for repainting within ½ hour and dries hard in 16 hours.

c. Certain exceptions to the regulations concerning painting exist. Fire-control instruments, sighting equipment, and other items which require a crystalline finish will not be painted with olive-drab enamel.

d. Complete information on painting is contained in TM 9-850.

28. Preparing for painting.—*a.* If the base coat on the matériel is in poor condition, it is more desirable to strip the old paint from the surface than to use sanding and touch-up methods. After stripping, it will then be necessary to apply a primer coat.

b. Primer, ground, synthetic, should be used on wood as a base coat for synthetic enamel. It may be applied either by brushing or spraying. It will brush satisfactorily as received or after the addition of not more than 5 percent by volume of thinner. It will be dry enough to touch in 30 minutes, and hard in 5 to 7 hours. For spraying, it may be thinned with not more than 15 percent by volume of thinner. Lacquers must not be applied to the primer, ground, synthetic, within less than 48 hours.

c. Primer, synthetic, rust inhibiting, for bare metal, should be used on metal as a base coat. Its use and application are similar to those outlined in *b* above.

d. The success of a job of painting depends partly on the selection of a suitable paint, but also largely upon the care used in preparing the surface prior to painting. All parts to be painted should be free from rust, dirt, grease, kerosene, oil, and alkali, and must be dry.

29. Painting metal surfaces.—If metal parts are in need of cleaning, they should be washed in a liquid solution consisting of ½ pound of soda ash in 8 quarts of warm water, or an equivalent solution, then rinsed in clear water and wiped thoroughly dry. Wood parts in need of cleaning should be treated in the same manner, but the alkaline solution must not be left on for more than a few minutes and the surfaces should be wiped dry as soon as they are washed clean. When artillery or automotive equipment is in fair condition and only marred in spots, the bad places should be touched with enamel, synthetic, olive-drab, lusterless, and permitted to dry. The whole surface will then be sandpapered with paper, flint, No. 1, and a finish coat of enamel, synthetic, olive-drab, lusterless, applied and allowed to dry thoroughly before the matériel is used. If the equipment is in bad condition, all parts should be thoroughly sanded with paper, flint, No. 2, or equivalent, given a coat of primer, ground, synthetic, and permitted to dry for at least 16 hours. They will then be sandpapered with paper, flint, No. 00, wiped free from dust and dirt, and a final coat of enamel, synthetic, olive-drab, lusterless, applied and allowed to dry thoroughly before the matériel is used.

30. Paint as camouflage.—Camouflage is now a major consideration in painting ordnance vehicles, with rust prevention secondary. The camouflage plan at present employed utilizes three factors: color, gloss, and stenciling. Vehicles are painted with enamel, synthetic, olive-drab, lusterless, which was chosen to blend in reasonably well with the average landscape.

31. Removing paint.—After repeated paintings, the paint may become so thick as to crack and scale off in places, presenting an unsightly appearance. If such is the case, remove the old paint by use of a lime and lye solution (see TM 9–850 for details) or remover, paint and varnish. It is important that every trace of lye or other paint remover be completely rinsed off and that the equipment be perfectly dry before repainting is attempted. It is preferable that the use of lye solutions be limited to iron or steel parts. If used on wood, the lye solution must not be allowed to remain on the surface for more than a minute before being thoroughly rinsed off and the surface wiped dry with rags. Crevices or cracks in wood should be filled with putty and the wood sandpapered before refinishing. The surfaces thus prepared should be painted according to directions in paragraph 29.

32. Painting lubricating devices.—Oil cups, grease fittings, oil-holes, and similar lubricating devices, as well as a circle about three-fourths of an inch in diameter at each point of lubrication, will be painted with enamel, red, water resisting, in order that they may be readily located.

Section VII

MATÉRIEL AFFECTED BY GAS

	Paragraph
Protective measures	33
Decontamination of matériel	34

33. Protective measures.—*a.* For matériel in constant danger of gas attacks, whether from chemical clouds or chemical shells, care should be taken to keep all unpainted metal parts of matériel, with the exception of ammunition, lightly coated with oil and protected with covers while not in use. Care must be taken that the oil does not come in contact with the optical parts of the instruments, with leather or canvas fittings, or with ammunition. The latter should be kept in sealed containers.

b. Ordinary fabrics offer practically no protection against mustard gas and lewisite. Rubber and oilcloth are penetrated if sufficient time is given. The greater the length of time allowed for penetration, the greater the danger of wearing these articles. For example, rubber boots which have been worn in an area contaminated with mustard gas may offer a grave danger to men who wear them several days after the bombardment. Impermeable clothing will resist penetration for over an hour, but should not be worn longer than that.

34. Decontamination of matériel.—*a. Cleaning.*—(1) All unpainted metal parts of matériel that have been exposed to any gas except mustard and lewisite must be cleaned as soon as possible with dry-cleaning solvent or denatured alcohol and wiped dry. Following this cleaning all parts should be coated with engine oil or sperm oil.

(2) In the event ammunition has been exposed to gas, it must be thoroughly cleaned before it can be fired. To clean ammunition, a noncorrosive decontaminating agent or, if this is not available, strong soap and cool water should be used. After cleaning, wipe all ammunition dry with clean rags. *Do not use dry powdered decontaminating agent (chloride of lime) on or near ammunition supplies*, as flaming occurs through the use of the chloride of lime on liquid mustard gas.

b. Decontamination.—The following measures should be taken for the removal of liquid chemicals (mustard, lewisite, etc.) from matériel. For all of these operations, a complete suit of impermeable clothing and a service gas mask must be worn. Immediately after the removal of the suit, a thorough bath with soap and water (preferably hot) must be taken. If any skin areas have come in contact with mustard, if even a very small drop of mustard gets into an eye, or if the vapor of mustard has been inhaled, it is imperative that complete first-aid measures be given within 20 or 30 minutes after exposure. First-aid instructions are given in TM 9–850 and in FM 21–40. Garments exposed to mustard gas can be decontaminated. If impermeable clothing has been exposed to vapor only, it may be decontaminated by hanging in the open air, preferably in the sunlight, for several days. It may also be cleaned by steaming for 2 hours. If the impermeable clothing has been contaminated with liquid mustard gas, steaming for 6 to 8 hours will be required. Various kinds of steaming devices can be improvised from materials available in the field.

(1) Commence by freeing matériel of dirt through the use of sticks, rags, etc., which must be buried immediately after this operation.

(2) If the surface of the matériel is coated with grease or heavy oil, this grease or oil should be removed before decontamination is undertaken. Dry-cleaning solvent or other available solvents for oil should be used with rags attached to ends of sticks. Following this, decontaminate the matériel with bleaching solution made by mixing one part decontaminating agent (chloride of lime) with one part water. This solution should be swabbed over all surfaces. Wash off with water, dry, and oil all surfaces.

(3) All unpainted metal parts and instruments exposed to mustard or lewisite gas must be decontaminated with a noncorrosive decontaminating agent, mixed one part solid to fifteen parts solvent (acetylene tetrachloride). If this is not available, use warm water and soap. Bleaching solution must not be used, because of its

corrosive action. Instrument lenses may be cleaned with lens tissue paper only, using a very small amount of ethyl alcohol. Coat all metal surfaces lightly with engine oil or sperm oil.

(4) In the event that a decontaminating agent (chloride of lime) is not available, matériel may be temporarily cleaned with large volumes of hot water. However, mustard lying in joints or in leather or canvas web is not removed by this procedure and thus will remain a constant source of danger until the matériel can be properly decontaminated. All mustard washed from matériel in this manner lies unchanged on the ground, necessitating that the contaminated area be plainly marked with warning signs before abandonment.

(5) The cleaning or decontaminating of matériel which has been contaminated with lewisite will wash arsenic compounds into the soil, poisoning water supplies in the locality for either men or animals.

(6) Leather or canvas web that has been contaminated should be scrubbed thoroughly with bleaching solution. In the event this treatment is insufficient it may be necessary to burn or bury such material.

c. In the case of vehicles that have been subjected to gas attack with the engine running, the air cleaners should be serviced by removing the oil, flushing with a dry-cleaning solvent, and refilling with the proper grade of oil. Instrument panels should be cleaned in the same manner as outlined for instruments. Seat cushions that are contaminated should be discarded. Washing the compartments thoroughly with bleaching solutions is the most that can be done in the field. Operators should be constantly on the alert, when running under conditions of high temperatures, for slow vaporization of the mustard or the lewisite. Exterior surfaces of vehicles should be decontaminated with bleaching solution. Repainting may be necessary after this operation.

d. Detailed information on decontamination will be found in FM 21–40 and TM 9–850.

Section VIII

ARMAMENT

	Paragraph
Gun mounts	35
Stabilizers	36
Sighting equipment	37

35. Gun mounts.—The 75-mm and 37-mm combination gun mounts, although given individual model numbers, are considered

TM 9-750
ORDNANCE MAINTENANCE

component parts of the tank. For more detailed instructions on operation, care, and preservation of the guns and their mounts, refer to the various manuals for the guns (see appendix). The ammunition stowage chart gives the location of ammunition. Also see figure 13.

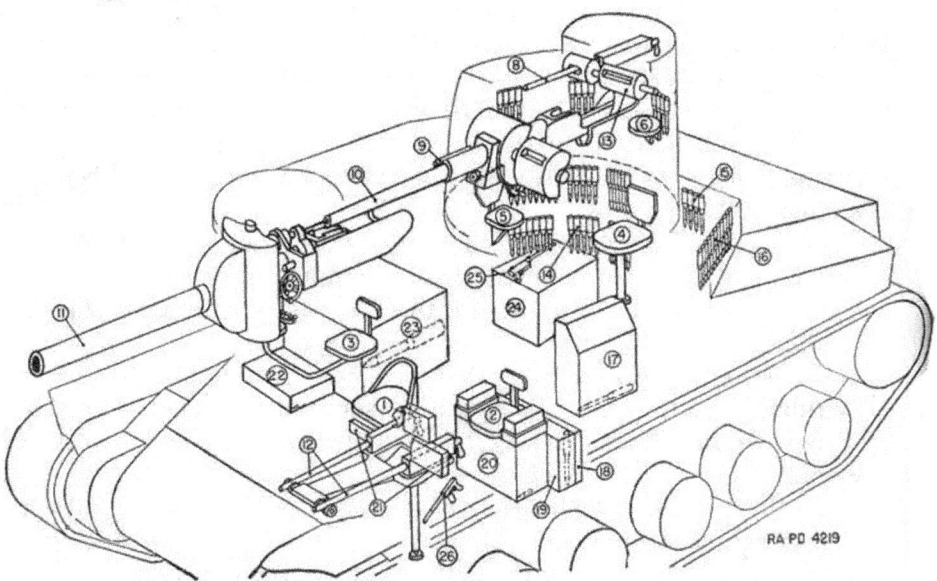

1. Driver's seat.
2. Radio operator's seat.
3. 75-mm gunner's seat.
4. 37-mm gunner's seat.
5. 37-mm gun loader's seat.
6. Tank commander's seat.
8. Cal. .30 machine gun.
9. Cal. .30 machine gun.
10. 37-mm gun.
11. 75-mm gun.
12. 2 cal. .30 machine guns.
13. Protectoscopes.
14. 51 rounds 37-mm ammunition carried in turret.
15. 13 rounds 37-mm ammunition.
16. 11 rounds 37-mm ammunition.
17. 42 rounds 37-mm ammunition.
18. Ten 100-round belts cal. .30 ammunition.
19. 20 rounds 37-mm ammunition.
20. Fourteen 250-round belts containing 225 rounds cal. .30 ammunition.
21. Two 250-round belts containing 225 rounds cal. .30 ammunition.
22. Twenty-five 100-round belts cal. .30 ammunition.
23. 41 rounds 75-mm ammunition; six 100-round belts cal. .30 ammunition.
24. 42 rounds 37-mm ammunition.
25. Submachine gun.
26. Submachine gun. Carried in tank but not shown on drawing are 9 rounds 75-mm ammunition carried in cartons and twenty-four 50-round clips cal. .45 ammunition.

FIGURE 13.—Armament.

a. 75-mm gun and mount.—(1) This gun mount is located in the right front of the crew compartment and mounts the 75-mm tank gun M2. The gun and mount are designed so as to provide protection to the tank personnel under all conditions of traverse and elevation.

MEDIUM TANKS M3, M3A1, AND M3A2

AMMUNITION STOWAGE CHART

Type	Number of rounds	Location
75-mm	41	Box on floor crew compartment, right side, directly behind 75-mm gun.
	9	Carried in cartons (portable).
	50	
37-mm	42	Box rear left corner (17, fig. 13).
	42	Box rear right corner (24, fig. 13).
	20	Box on floor, directly behind radio operator (19, fig. 13).
	51	3 rows mounted on racks on turret wall (14, fig. 13).
	13	Mounted on rack rear wall, fighting compartment (15, fig. 13).
	11	Above Homelite generator on rack rear left fighting compartment (16, fig. 13).
	179	
Cal. .45	20 50-round magazines	Rear right corner rack, top 37-mm box.
	2 50-round magazines	In guns.
	4 50-round magazines	Box, left of radio operator.
	1,300 rounds.	
Cal. .30	2 225-round belts [1]	In tray of cal. .30 gun feed box alongside gun (21, fig. 13).
	2 225-round belts [1]	Turret platform.
	1 225-round belt [1]	In combination mount.
	10 100-round belts	Box behind radio operator seat, with 37-mm supply (18, fig. 13).
	3 225-round belts [1]	Beside propeller shaft housing.
	6 100-round belts	Box over propeller shaft next to 75-mm supply.
	4 100-round belts	Above propeller shaft housing.
	14 225-round belts [1]	In trays under radio operator's seat (20, fig. 13).
	4 225-round belts [1]	Above propeller shaft (forward).
	6 225-round belts [1]	On tool box.
	32 225-round belts, 7,200 rounds	
	20 100-round belts, 2,000 rounds	
	9,200 rounds	

[1] These belts have a capacity of 250 rounds but are loaded with 225 rounds each.

TM 9-750
35   ORDNANCE MAINTENANCE

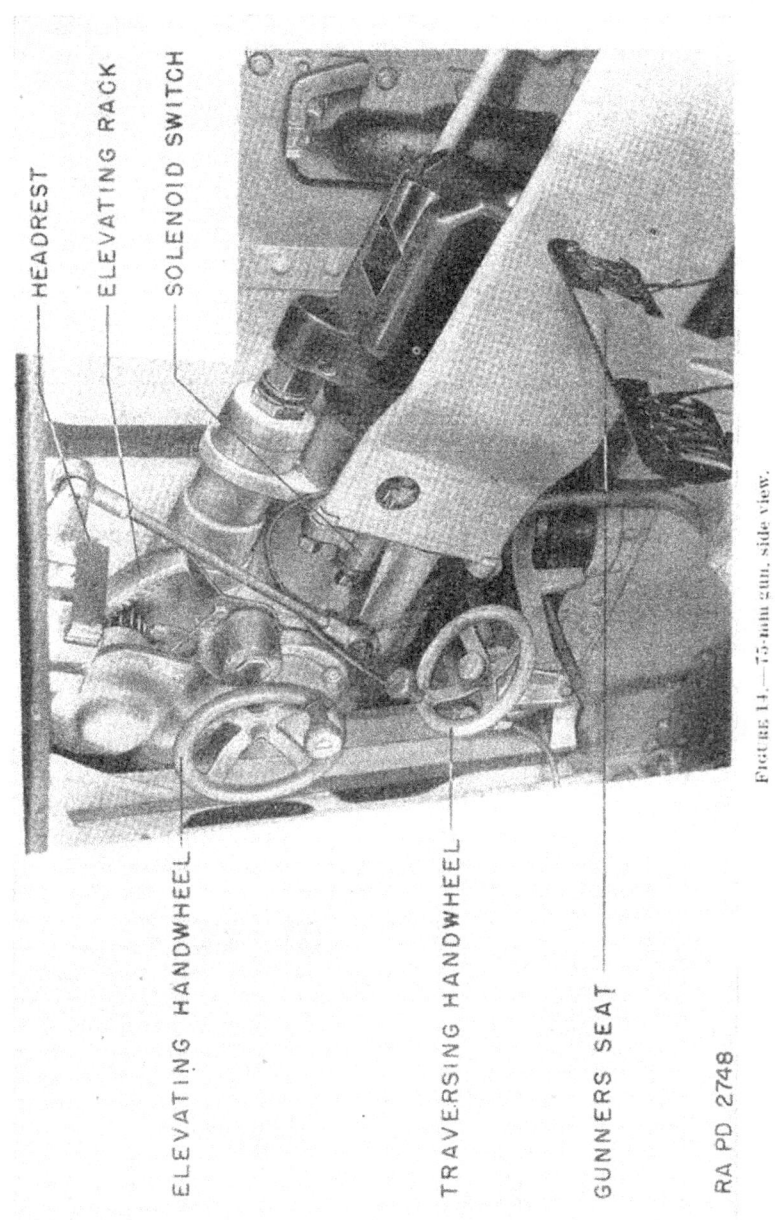

Figure 14.—75-mm gun, side view.

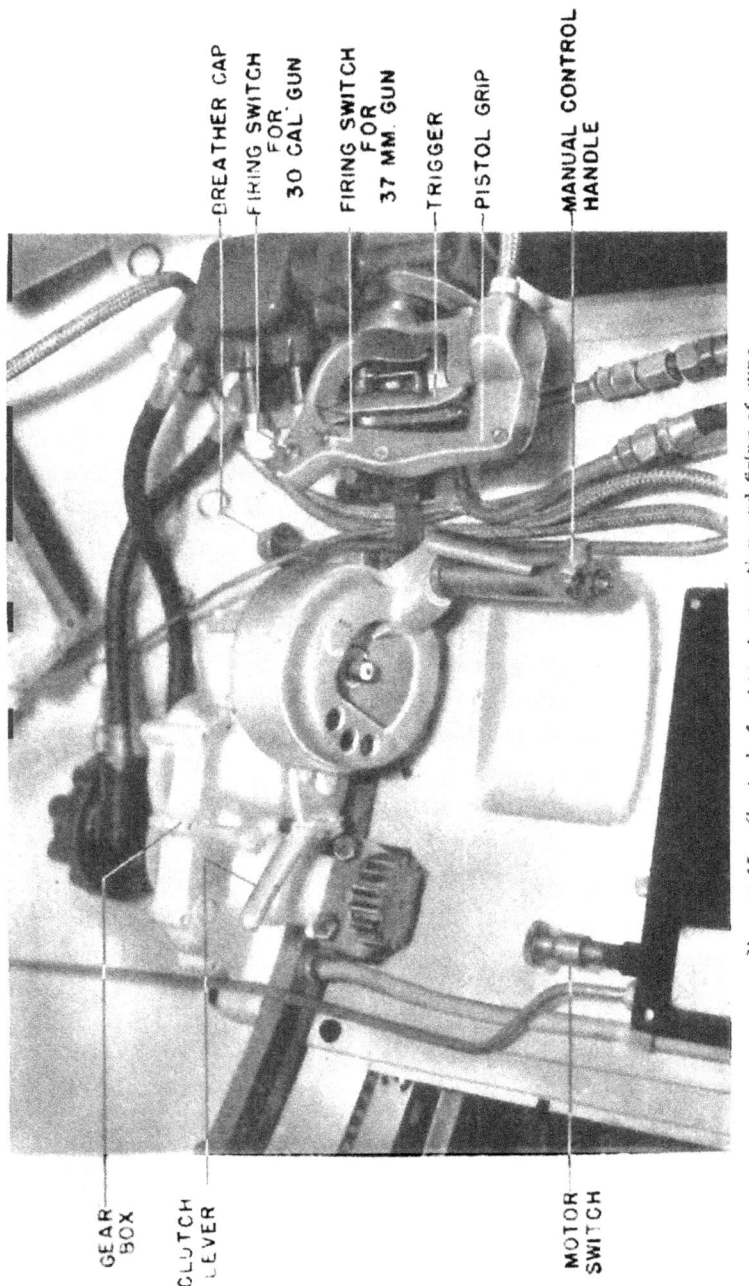

FIGURE 15.—Controls for turret operation and firing of guns.

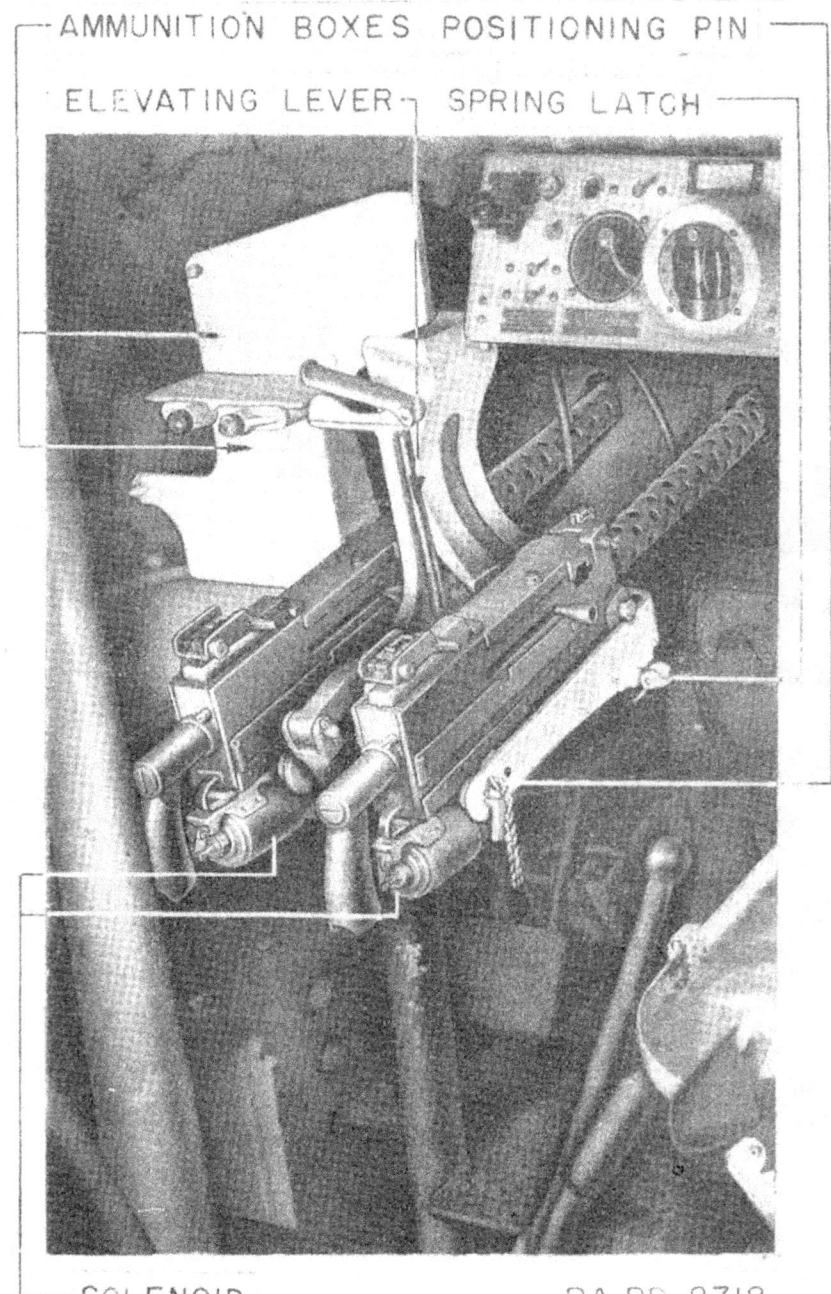

Figure 16.—Cal. .30 bow guns.

(2) The gunner sits in a seat on the left side of the gun mount. His position in relation to the gun does not change when the gun is elevated or traversed.

(3) The 75-mm gun is traversed 15° in each direction by the lower handwheel immediately in front of and facing the gunner. Clockwise turning of the handwheel traverses the gun toward the right; counterclockwise, toward the left.

(4) The gun may be elevated 20° or depressed 9° by turning the upper handwheel directly in front of the gunner. The wheel is turned clockwise for elevating and counterclockwise for depressing.

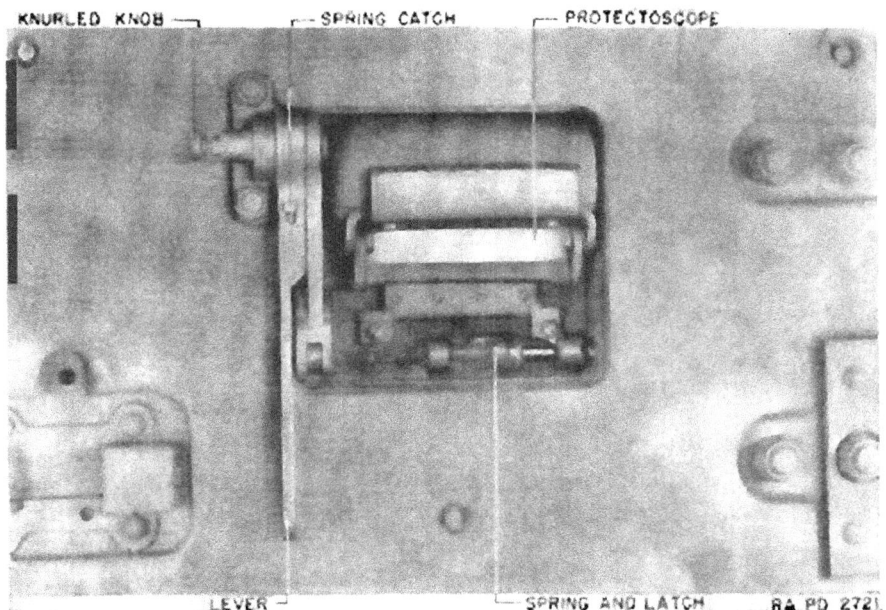

Figure 17.—Pistol port with protectoscope.

(5) The gunner sights through the periscope, into which is built a telescopic sight. An additional periscope and sight are carried in the tank. The entire sighting device may be removed, if damaged, by pulling out on the pin in the rear of the housing and pulling down on the ring attached to the bottom of the periscope. The telescope may then be removed by taking off the bracket which holds it in the vision piece. A new sighting device may be pushed into the housing if the catch pin is pulled outward to permit its entrance.

(6) In the rear of the central platform in the main fighting compartment is a space for the 75-mm gun loader. Close by him is

TM 9-750
35   ORDNANCE MAINTENANCE

hung one of the caliber .45 submachine guns M1928A1 for use through the pistol ports of the tank. Refer to figure 13 for ammunition stowage.

(7) The 75-mm gun can be fired manually or electrically. The gun is fired electrically by depressing the button located in front

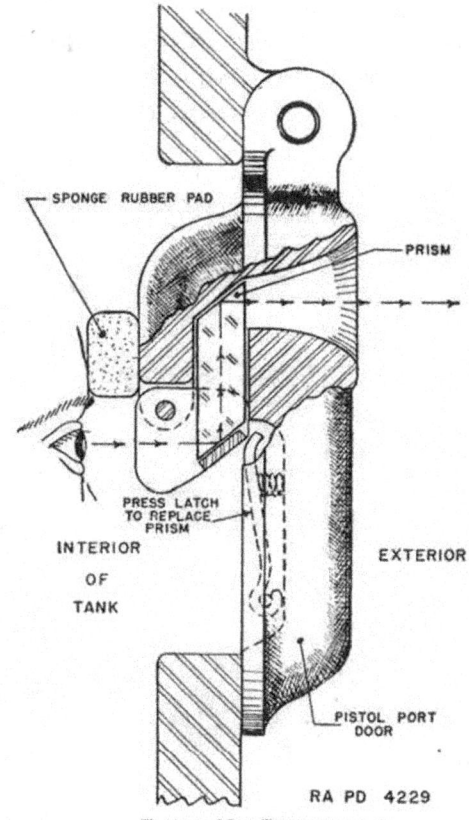

FIGURE 18.—Protectoscope.

of gunner, above handwheel. (See fig. 14.) The gun is fired manually by depressing a button located in the traversing mechanism.

b. Turret guns and mounts (fig. 15).—The guns located in the turret consist of a 37-mm gun M5 or M6 and a caliber .30 machine gun M1919A4, which are mounted together and move as one unit. The guns are traversed by turning the whole turret assembly, using the hydraulic system or by hand.

NOTE.—The gun tubes for the 37-mm guns M5 and M6 are not to be interchanged.

MEDIUM TANKS M3, M3A1, AND M3A2

For operation of turret traverse see section XIV, chapter 2. The elevating handwheel is located on the left side of the 37-mm gun directly in front of the gunner. Turning the handwheel counterclockwise depresses the guns a maximum of 7° and turning clockwise elevates them to a maximum of 60°. The guns are fired

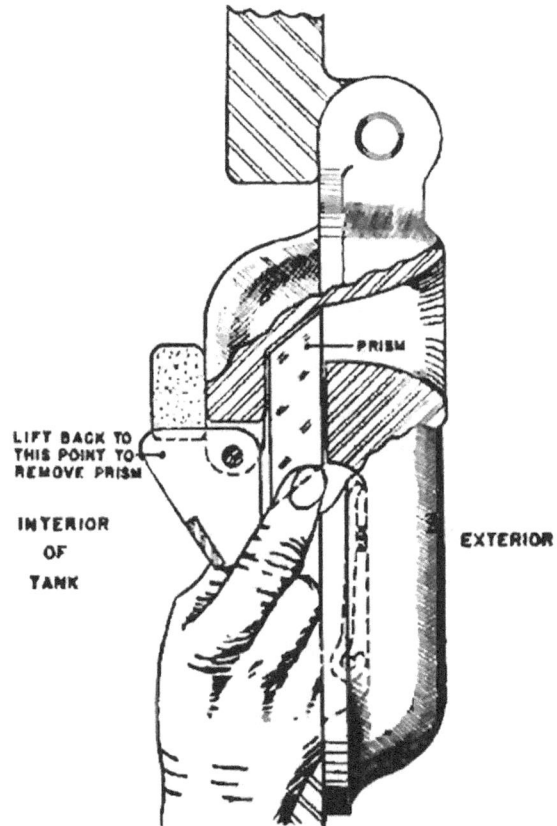

FIGURE 19.—Removal of prism from protectoscope.

electrically by squeezing the safety grip and depressing the switches on the hydraulic traversing control handle (fig. 15). The top switch operates the caliber .30 machine gun and the bottom switch operates the 37-mm gun. The gunner sits on the left side of the mount and the loader on the right. Care must be exercised to see that the loader's hand is not caught in the automatic breechblock. See figure 13 for ammunition stowage.

c. Cupola guns.—A caliber .30 machine gun M1919A4 is mounted in the cupola. The gun can be traversed 360° by rotating the cupola.

The gun may also be traversed to a limited extent independent of the cupola. Maximum elevation is 60° and maximum depression 8½°. The gun is fired manually by squeezing the trigger.

d. Bow guns (fig. 16).—The bow guns are located in the left side of the driver's compartment, projecting through the front plate, and consisting of two caliber .30 machine guns M1919A4 which are fired by the driver. The guns are elevated manually and have no traverse. Traverse is accomplished by steering the tank. The guns may be fired electrically by pressing switches mounted on the steering levers or manually by conventional triggers. Relay switches are located in junction box at left of instrument panel.

e. Tripods.—Tripods are carried on the outside of the tank for use with the bow machine guns.

f. Submachine guns.—Two Thompson caliber .45 submachine guns are normally carried in the main fighting compartment of the tank. These weapons may be used through the pistol ports or partially opened roof of the cupola, and when dismounting is necessary. Brackets for one submachine gun are located below the caliber .30 bow machine guns and to the left of the radio operator. The second submachine gun is mounted on brackets in the upper rear right-hand corner of the fighting compartment.

36. Stabilizers.—*a. Description.*—Both the 37-mm gun in the turret and the 75-mm gun in the right sponson are stabilized by equipment which automatically compensates for the pitching of the moving tank. The stabilizer tends to keep the gun in the same position in a vertical plane regardless of the position of the tank. The guns are stabilized by separate systems each consisting principally of an electric motor driven pump, piston and cylinder assembly, a control unit, an electric control box, recoil switch, and a master switch.

b. Controls.—The two stabilizer systems are similar and have the same type of controls. The following instructions will apply to both systems.

c. Starting the units.—To start either of the units proceed as follows:

(1) Set the stiffness control at zero.

(2) Disengage the hand elevating mechanism. This is accomplished in the case of the 37-mm system by pulling out the disengaging ratchet knob (see fig. 76) and pulling downward. This automatically closes the disengaging switch. The hand elevating mechanism for the 75-mm gun is disengaged by moving the clutch lever (fig. 77) which also closes the disengaging switch.

MEDIUM TANKS M3, M3A1, AND M3A2

(3) Turn the elevating handwheel until the control unit is approximately in a vertical position.

(4) Push the master control switch to the "on" position.

d. Operation.—The stabilizers should not be operated except when the tank is in motion. A red glow from the light on the control boxes signifies that the system is in operation.

(1) *Stiffness adjustment.*—After the system has warmed up turn the stiffness adjusting knob slowly in the direction of the arrow. When the gun begins to vibrate decrease the adjustment until the vibrating is eliminated. As a check, press on the breech suddenly and release. If the gun comes to rest almost immediately after a sharp displacement, it is properly adjusted. It may be necessary to change the stiffness adjustment from time to time as the viscosity of the oil changes.

(2) *Recoil adjustment.*—The recoil adjustment must be made by trial when firing the gun. Gradually turn the recoil adjusting knob in the direction of the arrow until the point is reached where the gun keeps its set angular position during recoil.

(3) *Performance.*—When the stabilizer equipment is running, the gun is elevated or depressed by simply turning the handwheel. This action changes the angular relation between the gun and the control unit and the gun automatically takes up a new desired position. If the stabilizer equipment is operating satisfactorily, it will keep the gun at a set angular position within certain limits while the tank is pitching. Therefore, when aiming, allow the stabilizer to control the position of the gun, and use the handwheel only when necessary. The operator must guard against continuing to turn the handwheel after the gun has reached its maximum limits of travel in the elevation plane. A continued turning of the handwheel with the gun against its stop will only displace the control unit from its vertical position and the result will be an overload on the tank battery.

e. Care and preservation.—(1) *Oil level.*—The oil levels in the oil reservoirs should be checked daily. The reservoir should be kept two-thirds full of hydraulic oil.

(2) *Lubrication.*—Both units are provided with alemite fittings (see fig. 80) and should be lubricated every 25 hours with chassis grease.

37. Sighting equipment.—*a.* The description, operation, adjustment, and care and preservation of the 75-mm gun sighting equipment are covered in TM 9-307.

b. The sighting equipment on the 37-mm combination gun mount is of the periscope type. This is used as an observing instrument

TM 9-750
37   ORDNANCE MAINTENANCE

and as a sighting element. The vision device is located at the eye level of the gunner as he sits in position. A telescope sight is built into the right side of the vision device. The entire vision device is readily removed by pressing in the lever button located in front of the upper left portion of the headrest and then pulling on the ring fastened to the bottom of the housing. The telescope sight can then be removed by taking off the bracket which fastens it into the periscope. Extra vision devices and telescopes are carried in conveniently located brackets so that rapid replacement can be made when necessary.

CHAPTER 2

ORGANIZATION INSTRUCTIONS

	Paragraphs
SECTION I. General information on maintenance	38–39
II. Equipment and special tools	40–41
III. Engine and accessories	42–60
IV. Fuel system	61–67
V. Cooling system	68–69
VI. Clutch	70–81
VII. Propeller shaft	82–84
VIII. Transmission, differential, and steering brakes	85–90
IX. Final drive	91–93
X. Tracks and suspensions	94–99
XI. Electrical equipment and instruments	100–118
XII. Auxiliary generating unit	119–121
XIII. Stabilizers	122–126
XIV. Turret	127–131
XV. Preparation for shipment and storage	132–134

SECTION I

GENERAL INFORMATION ON MAINTENANCE

	Paragraph
Scope	38
Engine trouble shooting	39

38. Scope.—*a.* The scope of maintenance and repairs by the crew and other units of the using arms is determined by the ease with which the project can be accomplished, the amount of time available, the nature of the terrain, weather conditions, temperatures, concealment, shelter, proximity to hostile fire, the equipment available, and the skill of the personnel. All of these are variable and no exact system of procedure can be prescribed or followed.

b. The following are the maintenance duties which may be performed by the using arm maintenance personnel. All other replacements and repairs will be performed by the ordnance maintenance personnel.

(1) *Engine R-975-EC2.*

Engine, service and remove (see par. 59).
Manifold, exhaust, and muffler, replace.
Rod, valve push, replace.

Valves, clearance, adjust.
Valves, check timing.
Valve, rocker, replace.
(2) *Oiling system.*
Cooler, replace.
Cooler bypass valve, clean, check, and replace.
Tank, replace.
Lines, external, clean and replace.
Strainers, clean.
Dilution valve, clean and replace.
Cuno filter, disassemble and clean.
Cuno filter, replace.
Oil pressure, adjust.
Pump, replace.
Temperature gage, replace.
Pressure gage, replace.
(3) *Cooling system.*
Fan cowling, tighten and clean.
(4) *Fuel system.*
Pressure regulating valve, check and replace.
Fuel cut-off valve, check and tighten.
Carburetor, replace.
Carburetor bowl, drain.
Carburetor elbow, replace.
Air cleaner, replace and service.
Gaskets, carburetor, to elbow to manifold, replace.
Lines, clean, repair, and replace.
Primer, lines, replace.
Primer pump, replace.
Pump, replace.
Filter, replace and service.
Tanks, clean and replace.
(5) *Electrical system.*
Magnetos, replace.
Magnetos, time to engine timing when installing magnetos.
Magneto points, adjust.
Ammeter, replace.
Voltmeter, replace.
Switch, magneto, replace.
Switch, battery, replace.
Switch, starter, magnetic, replace.
Fuel cut-off solenoid, replace.

Starter assembly, replace.
Starter solenoid, replace.
Generator assembly, replace.
Generator, check output.
Voltage regulator, replace.
Booster coil, replace.
Battery, charge, replace or service.
Siren, replace.
Fuse block assembly, replace.
Collector ring assembly, service and replace.
Auxiliary generator unit, service and replace.
Auxiliary generator, replace spark plug.
Wiring, replace.
Trace trouble with light.
Spark plugs, replace.
Harness, ignition wire assembly, replace.

(6) *Suspension.*
Bogie and idler wheels, replace.
Roller, track supporting assembly, replace.
Bogie components, replace.
Track, replace or reverse.
Track, components, replace.
Wheel bearings and oil seals, bogie or idler, replace.
Track, adjust.
Volute springs, replace.
Grousers, install and remove.

(7) *Hull and turret.*
Tighten or replace bolts, nuts, and screws.
Seats, replace.
Insulation, replace or reverse.
All pads, replace.
Turrent gun plate, replace.
Vision device, replace.
Pistol port doors and peephole covers, service and replace.
Mudguards, repair or replace.

(8) *Transmission and driving units.*
Clutch, adjust and service.
Clutch plates or bearings, replace.
Final drive assembly, replace.
Gear train assembly, replace.   (See note.)
Hubs or sprockets, replace.
Propeller shaft, service, replace.

Shift lever, replace.
Steering band, replace. (On latest models 2d echelon can replace the brake bands. On the older models it will be necessary to take apart the brake drum and differential housings in order to remove the brake bands. In this case ordnance personnel will perform the job.)
Steering brakes, adjust.
Steering lever, replace.
Transmission, replace. (See note.)

(9) *Miscellaneous.*
Painting.
Cleaning.
Homelite unit, service and replace.
Lubrication.
Hydraulic steering mechanism, refill oil reservoir.
Armament, service.
Tachometer and cable, replace.
Speedometer and cable, replace.

NOTE.—The using arm is authorized to remove and reinstall engine and power train assemblies, which includes the transmission and differential housings. However, the replacement of an engine with another engine or the replacement of a power train assembly with another power train assembly must not be done by the using arms unless authorization is received from ordnance maintenance personnel.

39. Engine trouble shooting.—Before starting any engine always clear the cylinders of hydrostatic lock by turning the engine over two complete engine revolutions by hand. If extra resistance is felt or the engine cannot be turned over, remove the spark plugs in the lower three cylinders to drain the trapped fluid. If after draining, the engine cannot be turned over, call ordnance personnel. If, on the other hand, the engine can be turned over by hand and the engine fails to start after repeated attempts, any one or combination of the following conditions may be the cause.

a. Engine fails to turn over when starter solenoid is pressed.—(1) First check the battery gravity.

(*a*) If reading is 1.150 or less, replace with fully charged batteries and have the discharged batteries recharged.

(*b*) If reading is approximately 1.280 the battery is fully charged.

(2) Next check the condition of battery cables.

(*a*) Examine battery terminals for corrosion and battery cable for shorted or broken sections. Clean corroded terminals and replace broken cables.

(*b*) Examine for loose connections and tighten clamps.

(3) Operate the starter switch to ascertain if it is functioning. Have one man listen at solenoid for click as button is depressed. If no click is heard, replace the solenoid or disconnect the battery-starter motor cable at the starter motor, and while another person operates the starter switch, hold the cable against the housing. A spark should occur. If it doesn't, replace the solenoid.

(4) If the starter does not operate after solenoid clicks, starter should be replaced.

b. Engine turns but does not start.—(1) Check the ignition system.

(*a*) Check the ignition wires for breaks, worn sections, and loose connections. Remove a wire from the spark plug and check for length of spark.

(*b*) If the length of spark exceeds $3/8$ inch, it is an indication that the booster coil and magnetos are functioning properly.

(*c*) Next remove an upper and lower cylinder spark plug and examine for broken porcelain and corroded, burnt, or dirty electrodes. If dirty, clean and adjust all the spark plugs. Replace all broken or burnt spark plugs.

(*d*) If the spark is not of sufficient length, then check the booster coil and magneto.

    1. Inspect the booster coil by listening for a buzzing sound while the operator presses the starter switch. If a buzz is not heard, replace the coil.

    2. Check the output of the booster coil by disconnecting the high tension line to the right magneto at the magneto and turn the engine over by starter motor. The spark which occurs should be approximately $3/8$ inch long. If the spark is shorter than $3/8$ inch replace the coil. *Do not* attempt to adjust the spark.

    3. Check the magneto contact points (gap 0.012 inch). Inspect the points for pitting or burning. If the points are pitted or burnt, it may be due to a loose or defective condenser or overlubrication of the magneto. If the points are pitted slightly, the points may be smoothed with a fine file or abrasive stick. (Adjust gap.) If the points are badly pitted or burnt replace the magneto. *Do not* attempt to replace the points. If magneto does not function after preceding check, replace magneto.

(2) Check the fuel system.

(*a*) Check the amount of fuel in the fuel tanks.

(*b*) Check to determine if the fuel valves are open.

(*c*) Examine the carburetor for flooding. (Carburetor will be wet and leaking.) Flooding is usually due to sticking of the float or leakage of the float valve. If such is the case replace the carburetor.

(*d*) Check the fuel flow.

    1. Disconnect the inlet line at the carburetor, turn the engine over by using the starting motor, and watch for fluid flow.

    2. If no fluid flows, disconnect the carburetor inlet line at the fuel pump and blow through the line to determine if it is plugged. If it is plugged, clean by blowing compressed air through the line.

    3. If the inlet line is open, the trouble may be in the fuel pump. Test as follows: Disconnect the inlet fuel line to the fuel pump at the pump and if the line is open, fluid should flow. If fluid does not flow, the lines are plugged. Clean the fuel lines. If fuel does flow, reconnect the line to the pump. Now turn the engine over by starter motor; if no fluid appears at the discharge side of the pump, replace it; if fuel flows, check the pressure of fuel being delivered by the pump (3 pounds).

(3) After performing *all* the checks enumerated, if the engine will not start, substitute a new carburetor and attempt a start. If the engine will not start then remount the *original* carburetor and call the ordnance personnel.

Section II

EQUIPMENT AND SPECIAL TOOLS

	Paragraph
General	40
Fire extinguishing system	41

40. General.—The items listed below are carried with the vehicle and are strapped in place on the exterior of the hull or located inside the vehicle in the most convenient places available.

List of equipment	*Where carried*
Ax, chopping, single bit, 5 lb	On tank.
Belt, safety	In tank.
Book, Ordnance Motor (O. O. Form No. 7255).	In tank.
Cable, towing	On tank.

List of equipment	Where carried
Cover, headlight (sun glare)	On tank.
Cushions (2)	1 in tank; 1 in turret.
Extinguishers, fire, carbon dioxide, 4-lb., filled (2).	1 in tank; 1 in turret.
Extinguisher, fire, carbon dioxide, 10-lb., filled (for medium tank M3).	In tank under turret.
Goggles, automobile, amber, pair (QMC issue).	In tank.
Grousers (for medium tank M3)	In tank.
Helmet, tank	In tank.
Lamp, inspection M. T. M3, with bulb	In tank.
Lever, front door (for medium tank M3)	In tank.
Mattock, pick, M1 (5-lb. head)	On tank.
Padlock, 1½ in., keyed individually, 2 keys	On tank.
Paulin, 12 by 12 ft	In tank.
Shovel, D-handle, round point	On tank.
Strap, leather, type D (1¼ in.), 13 in. long (one each for ax, mattock handle, mattock head, and wrench).	On tank.
Strap, leather, type D (1¼ in.), 25 in. long (for tripod).	On tank.
Windshield (for medium tank M3)	In tank.

A tool kit (No. 3) is included with each tank.

41. Fire extinguishing system.—*a. Operation.*—(1) The fire extinguishing system is entirely manual in operation. It is therefore imperative that there be as little delay as possible in discharging the gas, as its effectiveness is materially increased by catching the fire in the beginning.

(2) To operate the system, pull one control handle. Control handles are located inside the tank at the turret basket bracket, adjacent to the fuel gages, and outside tank above the motor compartment. For second fire prior to recharging of first cylinder, pull other control handle.

NOTE.—The gas can also be released at cylinders by removing pull-out pin in control head and rotating manual lever.

b. Description.—(1) The fire extinguishing system consists of a supply of carbon dioxide gas stored in two steel cylinders, each having a capacity of 10 pounds of carbon dioxide; manual release gear for independent operation of each cylinder; and tubing to convey the gas to the engine compartment. The tubing is terminated in shielded discharge nozzles which effectively distribute the gas.

(2) The two gas cylinders are connected to the supply tubing by means of a double-check tee, which tee prevents the loss of gas into the crew compartment should one cylinder be operated while the other cylinder is removed for weighing or recharging.

(3) In addition to the built-in system for the engine compartment described above, two 4-pound portable extinguishers are provided for small fires in or about the tank. Full operating instructions are found on extinguisher name plate.

c. Principle.—(1) The fire extinguishing system uses carbon dioxide as the extinguishing agent. Carbon dioxide (not to be confused with carbon monoxide) is not poisonous but is suffocating.

(2) "Fast" fires, such as those involving gasoline or oil, are quickly extinguished by flooding the area with carbon dioxide gas. This reduces the oxygen content and creates an inert atmosphere which smothers the fire. "Slow" or "deep seated" fires, such as fires in baled cotton and similar substances, are extinguished by prolonged action of a high concentration of carbon dioxide. In addition to its smothering action, carbon dioxide is aided in extinguishing fire by its cooling effect.

(3) Since a person cannot breathe, but will suffocate in an atmosphere of carbon dioxide, caution must be taken before entering any space filled with this gas. Thoroughly ventilate the space into which the gas has been discharged to make certain that all portions contain only fresh air. Should it be necessary for a person to enter a space before it is thoroughly ventilated, he may do so for a short period by holding his breath.

(4) Should a person be overcome by carbon dioxide, it is essential that he be rescued from the space containing the gas within 5 minutes. To revive a person so overcome, give him plenty of fresh air and apply artificial respiration as in the case of drowning.

d. Maintenance.—After a fire, restore the system to its normal ready-to-operate condition as follows:

(1) Remove discharged cylinder as follows:

(*a*) Disconnect control head from cylinder valve by turning swivel nut (right-hand thread); raise clear of cylinder valve and support control head and cable tubing in approximately normal position.

(*b*) Remove connecting tube between double-check tee and valve outlet.

Caution.—Never remove cylinder with this connecting tube attached to cylinder valve outlet.

(*c*) Remove cylinder clamps; then remove cylinders.

MEDIUM TANKS M3, M3A1, AND M3A2

(2) While control head is disassembled from cylinder, remove cover exposing the cam (see fig. 20). Check cable clamp set screws to make certain cable does not pull out of clamp. In order to tighten

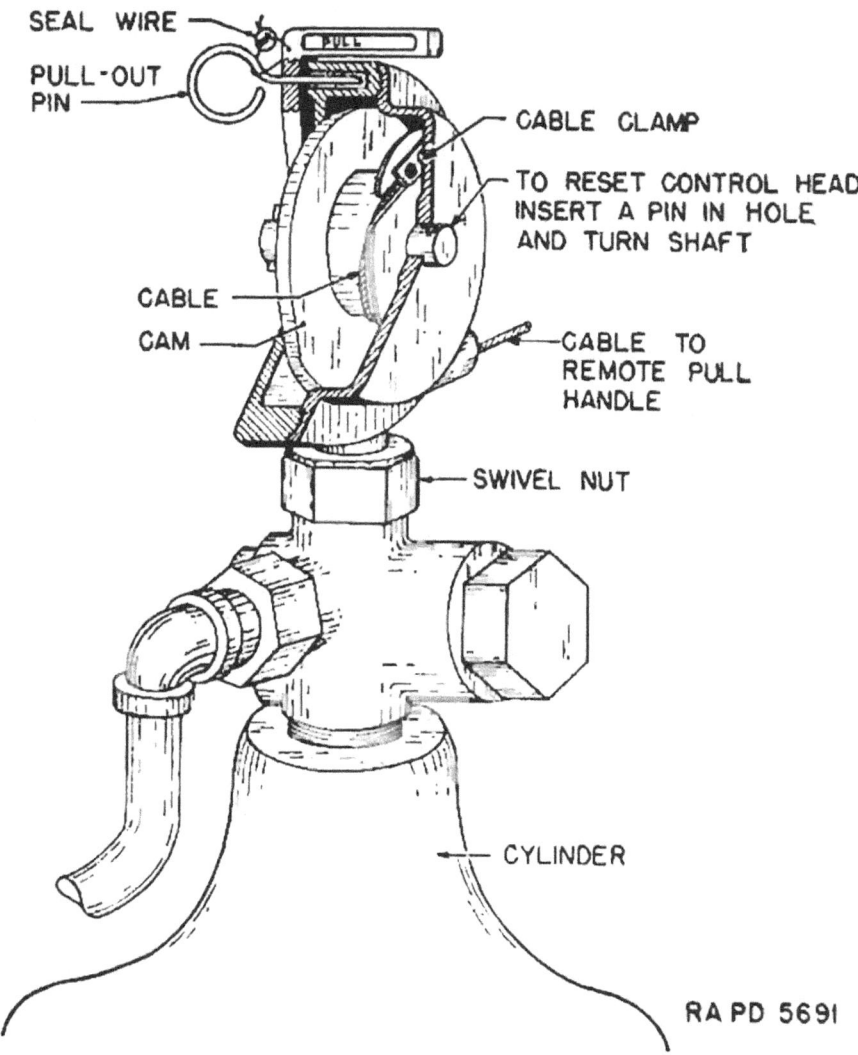

Figure 20.—Fire extinguisher system control head and cylinder valve.

set screws, a screw driver with narrow head must be used to make certain screw driver follows screw into hole.

(3) Reset remote control handle. (Control handles are located at the turret basket bracket adjacent to the fuel gages, and outside tank above the motor compartment.) Reset control head by insert-

TM 9-750
ORDNANCE MAINTENANCE

ing a pin or nail in control head shaft and turning counterclockwise (looking at manual lever side of control head) until clutch pin and arrow are lined up as shown in fig 21. If system has been operated at cylinder, replace pull-out pin and seal wire.

(4) Inspect the orifices in the shielded nozzles to make certain they are clear of all foreign matter.

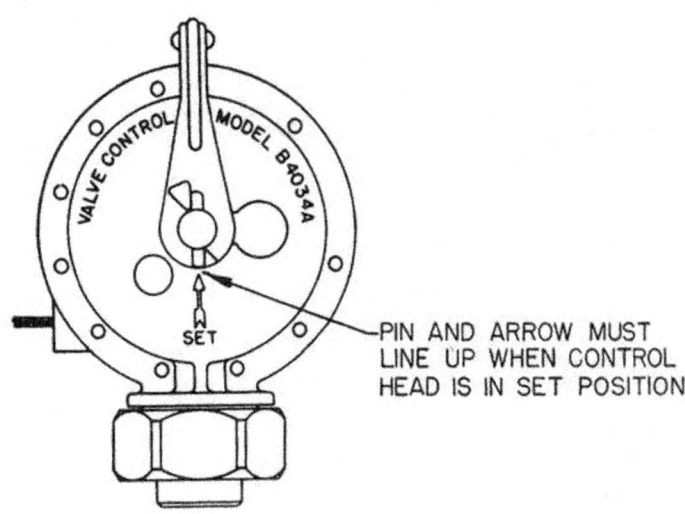

FIGURE 21.—Fire extinguisher system control head.

(5) Reinstall the fully charged cylinder as follows:

(a) Set cylinder in place and tighten clamps handtight to allow for any turning of cylinder that may be required in connecting tubing.

(b) Replace connecting tube between double-check tee and valve outlet.

(c) Tighten cylinder clamp securely.

(d) *Make certain* control head is in set position (pin lined up with arrow as shown in fig. 21) before reinstallation. Insert control head in cylinder valve, and before tightening swivel nut, check to see that control head has remained in "set" position. Then tighten swivel nut securely.

e. Periodic inspections.—The fire extinguishing system requires no more than ordinary care to insure its proper operation. As the system is for emergency use, it must be kept in operating condition at all times; therefore, frequent inspection should be made to insure that apparatus is intact. Check red cap on safety outlet of valve. If not intact, cylinder has been prematurely discharged due to high temperature and must be recharged immediately. The following inspections will be performed:

(1) *25-hour inspection.*—Inspect entire system for any mechanical damage. Make certain that shielded nozzles are free of all foreign matter.

(2) *50-hour inspection.*—Weigh cylinders to determine the carbon dioxide content. Do not attempt to determine content by means of a pressure gage. Empty weight and carbon dioxide charge are stamped on cylinder valve body. Proceed as follows:

(*a*) Remove cylinders as described in paragraph 41d(1).

(*b*) Weigh cylinders and subtract from this weight the empty weight that is stamped on valve body. Empty weight includes cylinder valve and cylinder, but does not include the control head. If the resulting net weight of either cylinder has decreased to below 9 pounds, cylinder must be recharged to its full rated capacity of 10 pounds.

(*c*) While both control heads are disassembled from cylinders, remove the cover exposing the cam (see fig. 20). Check cable clamp set screws to make certain that cable does not pull out of clamp. In order to tighten set screws, a screw driver with narrow head must be used to make certain screw driver follows screw into hole. After replacing cover, pull remote control handle to make certain that cable does not bind. The cam inside control head should rotate, and the pin advance.

(*d*) Reinstall cylinders as described in paragraph 41d(5).

TM 9-750

ORDNANCE MAINTENANCE

Section III

ENGINE AND ACCESSORIES

	Paragraph
General description	42
Construction and installation	43
Characteristics	44
Accessory case and drive mechanism	45
Oil pumps	46
Manifolds	47
Ignition shielding harness	48
Magnetos	49
Spark plugs	50
Booster coil	51
Starter	52
Generator	53
Fuel pump	54
Carburetor	55
Air cleaners	56
Governor	57
Engine lubrication and lubricants	58
Replacement of engine	59
Valve adjustments	60

42. **General description** (fig. 22).—*a.* The Wright Whirlwind R-975-EC2, a single row, air-cooled, radial, aviation type engine, is used.

(1) *Direction of rotation* (from rear or magneto end).

Crankshaft	Clockwise
Tachometer drive	Counterclockwise
Fuel pump	Do.
Starter	Do.
Generator	Do.
Magneto	Do.

(2) *Circuit gaps.*

Magneto breaker points gap	0.012 inch
Spark plug gap, mica plugs	0.012 inch

(3) *Engine oil.*

Pressure *should* be within following limits:

Minimum	50 lb.
Maximum	80 lb.
Quantity	36 qts.

Temperature at inlet:

Maximum allowable, degrees Fahrenheit	190
Desired, degrees Fahrenheit	140

MEDIUM TANKS M3, M3A1, AND M3A2

TM 9-750
42-43

b. The two magnetos, starter, generator, and fuel pump are mounted on the rear face of the rear crankcase section (fig. 22). The carburetor is mounted on the extreme lower end of the rear section.

c. In this section the flywheel end of the engine will be referred to as the "front" of the engine and the other end will be referred to as the "rear" of the engine. The terms "right" and "left" are used with reference to the engine as viewed from the rear. Horizontal and vertical positions of the engine will be referred to with respect

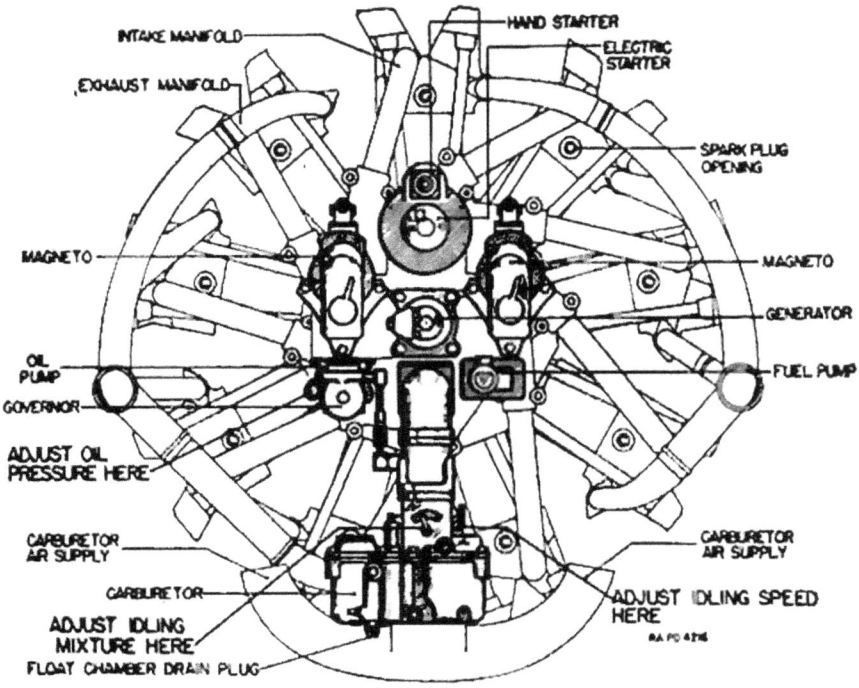

FIGURE 22.—Engine, showing accessories, adjustment of carburetor, and oil pressure.

to the position of the crankshaft. Direction of rotation is determined by viewing from the rear of engine, or in the case of side drives by looking toward the crankshaft.

43. Construction and installation.—*a. Construction.*—The nine cylinders of the engine are radially mounted on the crankcase and equally spaced. A formed, shielded conduit or harness contains all the wiring for the engine, including the high-tension cables. Baffles are installed over and between the finned cylinders for cooling.

b. Installation.—The engine is mounted in the tank as shown in figure 28.

44. Characteristics.

Make and type	Wright, radial, air-cooled.
Model and series	R-975-EC2.
Over-all diameter	45 inches.
Weight including accessories	1,370 lb.
Horsepower	400.
Number of cylinders	9.
Clearance between valve stem and rocker roller, cold	0.010 inch.
Firing order (clockwise viewed at magneto end)	1-3-5-7-9-2-4-6-8.
Location of No. 1 cylinder	Top.

45. Accessory case and drive mechanism.—The accessory case is attached to the rear face of the crankcase and houses the gear train that drives the various accessories, such as the magnetos, starter, generator, etc.

46. Oil pumps (fig. 22).—*a.* The supply of oil in the engine crankcase is maintained by oil pumps which draw oil from the engine oil tanks located at the bulkhead between the engine and fighting compartments, or which circulate the oil within the engine itself. These are gear type pumps which are supported on the lower left side of the rear crankcase section in the case of the main pump, and in the nose section in the case of the other scavenge pump. The main pump is divided into two sections: a separate pressure pump and a scavenge pump. This scavenge pump returns the oil from the upper section of the sump to the oil filter. The scavenge pump located in the nose section removes oil from the lower section of the sump. Both return lines from the scavenge pumps join together before running into the Cuno filter. The main oil pump incorporates a pressure relief valve which comprises a spring-loaded ball that rises when the desired pressure is reached and allows the excess oil to be drained into the inlet side of the pump. Pressure regulation is accomplished by turning the adjusting screw clockwise to raise the pressure and counterclockwise to lower the pressure which raises or lowers the tension in the spring (see fig. 22). The oil pressure should be between 50 and 80 pounds per square inch in normal operation.

b. Replacement.—To replace the oil pump assembly—

(1) Disconnect the inlet line at connector and outlet line at **Y** connection.

(2) Disconnect governor linkage at carburetor connection.

(3) Remove oil drain at governor.

(4) Remove the safety wire, stud nut, and plain washer from the eight studs on the main pump casting and remove the pump from the section of the crankcase. Caution must be exercised to keep foreign substances out of the crankcase during the operation. Separate oil pump and governor.

(5) To install a new pump, proceed in the reverse order, being careful to line up the drive gear properly.

c. To clean the strainer in the oil pump, remove the temperature bulb by removing the nut holding the bulb in the recess. Remove and clean strainer cover, spring, and strainer. In replacing the strainer be sure that the gasket is in good condition and centered properly in its recess.

47. Manifolds.—*a. Intake.*—The intake pipes are fastened to the intake port connections of each cylinder by means of a flange and three cap screws. The inner end of each intake pipe fits into the diffuser section of the crankcase and is sealed in place by a rubber seal and packing unit.

b. Exhaust.—The exhaust manifold is composed of two sections, one for the right side and one for the left side of the engine. The various branch pipes are mounted rigidly at the exhaust port flanges of the cylinder. The two outlets pass through the rear hull plates to the mufflers.

c. Supercharger.—The supercharger is used to force fuel mixture under pressure into the intake ports of the cylinder.

d. Manifold and muffler removal.—Certain sections of the manifolds may be removed with the engine remaining in the tank. The complete manifold cannot be removed with the engine in the tank. The following procedure is used to remove the entire muffler assembly:

(1) Remove rear top plate.

(2) Remove mufflers and elbows by removing 4 bolts on each mounting elbow. The muffler can be disconnected from the engine exhaust manifold by unclamping the manifold connection and pushing the pipe on the manifold side up into the muffler in order to clear the two ends (see fig. 22).

48. Ignition shielding harness.—All parts of the ignition wiring system are shielded to prevent radio interference. The shielding is constructed so that if any part is damaged it may be readily replaced without discarding the entire shielding structure. The ends of the conduits are equipped with an elbow which attaches directly to the shielded spark plug and facilitates easy removal. The ignition harness should be inspected at regular intervals to see that all the joints are tight and that the conduits are not crushed or broken.

TM 9-750
ORDNANCE MAINTENANCE

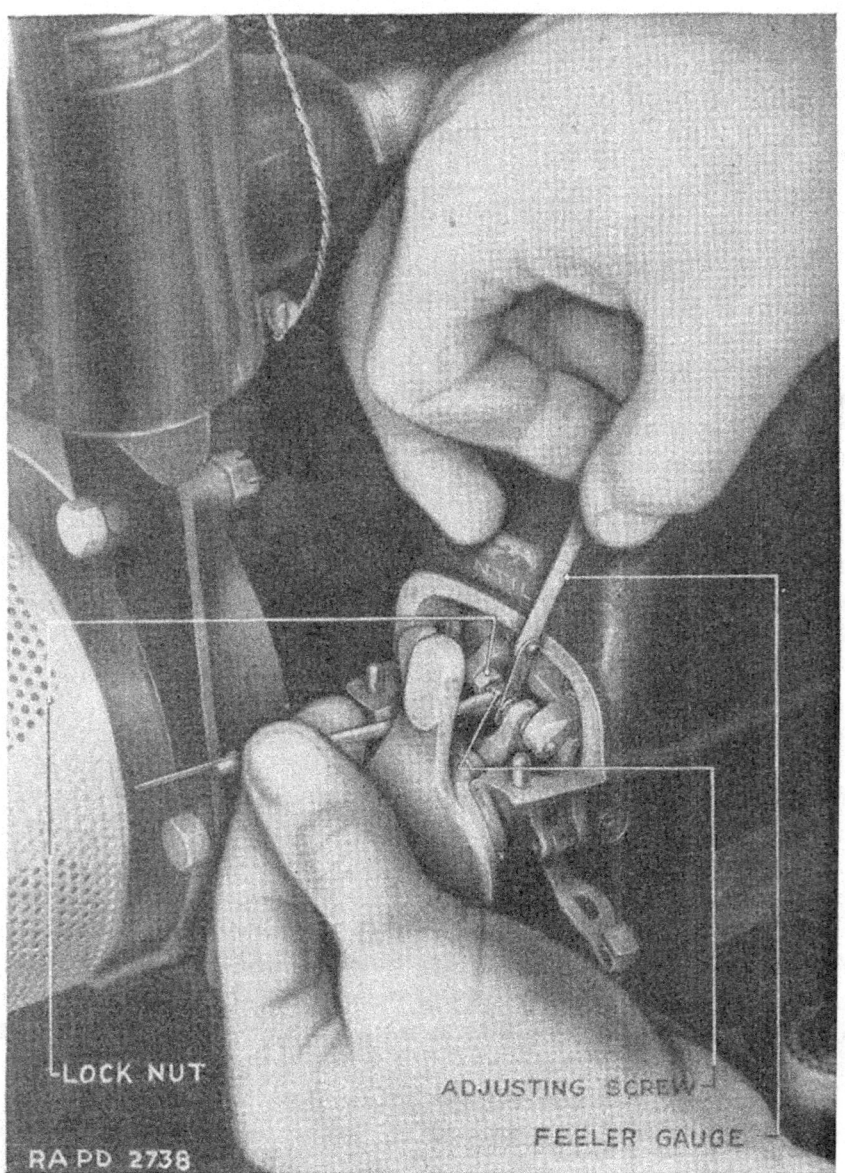

FIGURE 23.—Magneto point adjustment.

49. Magnetos.—*a. Description.*—Dual ignition is furnished by two magnetos mounted on the accessory case. The right magneto fires the front spark plugs and the left magneto fires the rear plugs. The magnetos are equipped with automatic spark advance, are shielded to prevent radio interference, and are flange mounted.

b. Adjustment (fig. 23).—Remove the breaker cover by releasing the safety ring on each side. Using a feeler gage, measure the clearance of the contacts when held wide open by the cam. This should be from 0.010 inch to 0.014 inch with 0.012 inch desired. To get the cam in a position to indicate the maximum gap, the engine will have to be turned over with the hand crank on the starter mechanism. To adjust the clearance between the contact points on the magneto, loosen the lock nut on the long contact screw, and turn the screw until there is 0.012-inch clearance between the contacts. Check clearance on all cam lobe positions.

NOTE.—The numbers on the magneto ends of the ignition wires correspond to the order in which they fire the spark plugs, and not to the cylinder numbers. Thus the wire to cylinder No. 3 is stamped "2" as it is the second in the cylinder firing order. (Cylinder firing order, clockwise viewed at magneto end, is 1-3-5-7-9-2-4-6-8.)

c. Maintenance.—Disassembly of the magneto other than the operation given herein will be done only by ordnance maintenance personnel.

d. Lubrication (see par. 18).—Each magneto has two oilholes, one at the top of the front end plate and the other on the main magneto cover. The main magneto cover oiler lubricates the breaker end ball bearing of the rotating magnet. Every 400 hours of operation, if the magneto is not replaced, apply 15 drops of oil in the front plate oiler. The breaker and oiler on the main cover should receive only three to five drops, as excess oil will foul the breaker contact points. Examine the felt wick at the bottom of the contact breaker to make sure it is saturated with oil. If oil appears on the surface of the felt when it is squeezed with the fingers, no additional lubricant is needed. If it is dry, however, saturate it with oil, engine, SAE 30.

e. Removal and replacement (fig. 24).—(1) Remove the two ¼-inch bolts at the top of the magneto cover and two safetypins. Loosen the two spring clamps at bottom and remove cover.

(2) Unscrew two terminal contacts at top of right magneto and pull out booster coil wire and ground wire. There is no booster coil wire on the left magneto.

TM 9-750
49-50   ORDNANCE MAINTENANCE

(3) Remove the two distributing blocks and rotate the engine with the hand crank until the two marks on the rotor gear coincide with the two marks on the magneto housing.

(4) Remove seal wire from the three nuts which hold magneto in place.

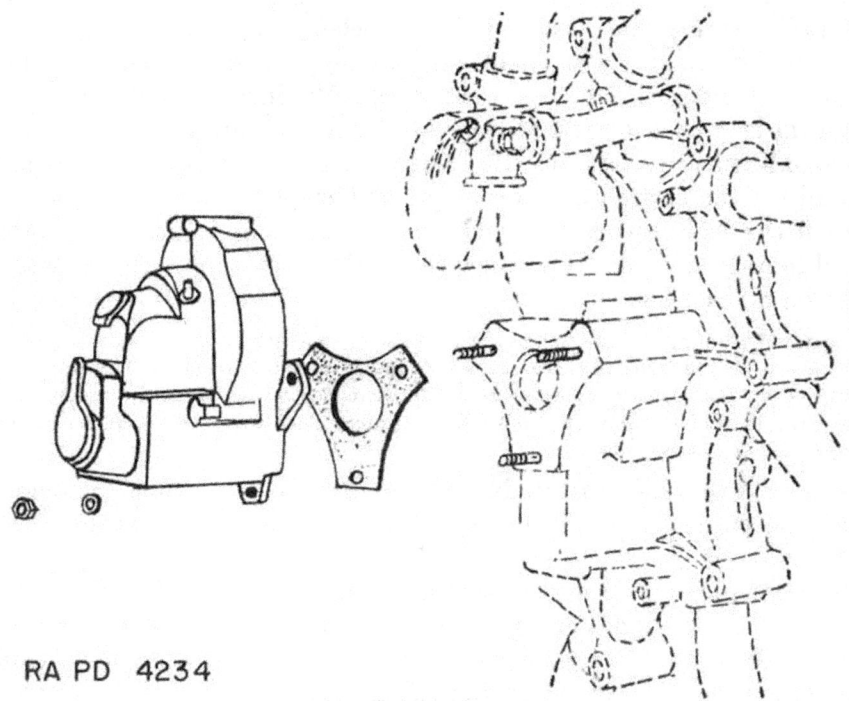

RA PD 4234

FIGURE 24.—Replacement of magneto.

(5) Remove the three nuts holding magneto to accessory case.

NOTE.—Care must be taken that the crankshaft is not turned until the new magneto is installed.

(6) Remove magneto.

(7) In replacing the magneto put a few drops of engine oil on the magneto coupling and turn the magneto until the two marks on rotor gear coincide with two marks on the magneto housing. The magneto may be lined up by tapping to each side. Continue with the reverse of the above steps.

50. Spark plugs (fig. 25).—*a.* The spark plugs used on this engine are the regular aircraft type, BG-314GS modified, and are radio shielded.

b. Replacement and removal.—(1) At the time of the 50-hour check, or whenever spark plug replacement is found necessary, the unserviceable plugs will be replaced by serviceable plugs.

(2) *To remove the spark plugs from the engine.*—(*a*) Remove the front and rear top plates over the engine.

RA PD 2756

FIGURE 25.—Spark plug for Wright engine.

(*b*) Remove the cable connection with a ⅝-inch open-end wrench. Remove the spark plug from the cylinder recess with a spark plug wrench (1-inch wrench No. 140).

(*c*) Remove inspection plate from bottom of hull under engine compartment.

(*d*) To facilitate removal of front spark plugs, remove inspection cover plates on fan cowling.

(*e*) The use of a socket wrench with a universal joint extension facilitates removal of certain plugs the location of which makes removal difficult with an ordinary socket wrench.

(*f*) Five plugs may be removed from the front and five from the rear of the engine by reaching down from the top of the com-

TM 9-750
50-51   ORDNANCE MAINTENANCE

partment. The remaining plugs can be reached through the opening in the hull beneath the engine.

(3) When installing spark plugs in the engine do not use wrenches with handles more than 10 inches long. It is possible to distort certain sections of the plug if too much force is used in tightening the plugs in the engine cylinders. Solid copper gaskets are to be used. When spark plugs are replaced the threads should be coated with antiseize compound to prevent "freezing" of the plugs.

Figure 26. Electric starter.

51. Booster coil.—*a. Description.*—A booster coil is provided which supplies an intense auxiliary spark across the points of the spark plugs to facilitate engine starting. The booster coil switch is located on the instrument panel, and is used to connect the booster coil at starting.

b. Replacement.—At the time of the 100-hour check, or whenever booster coil replacement is found necessary, the unserviceable unit will be replaced by a serviceable one. Disconnect cables with shielded connection to booster coil, remove bolts holding coil to bracket, and replace with new coil. Disassembly, servicing, and adjustment of the coil will be done only by ordnance maintenance personnel.

52. Starter (fig. 26).—*a. Description.*—A 24-volt direct electric starter is used on this engine. The starter should be used only for cranking the engine. It must not be used to move the tank when the engine is not functioning.

b. Hand cranking.—Provision is made at the top of the starter unit for hand cranking.

c. Electric cranking.—A starter switch is located on the instrument panel.

d. Starter.—If the starter fails to operate properly it should be replaced with a serviceable starter. The following procedure gives the method for removal and replacement of the starter.

(1) First open the battery switch.

(2) Remove the terminal guard cover and disconnect the cable from the starter switch.

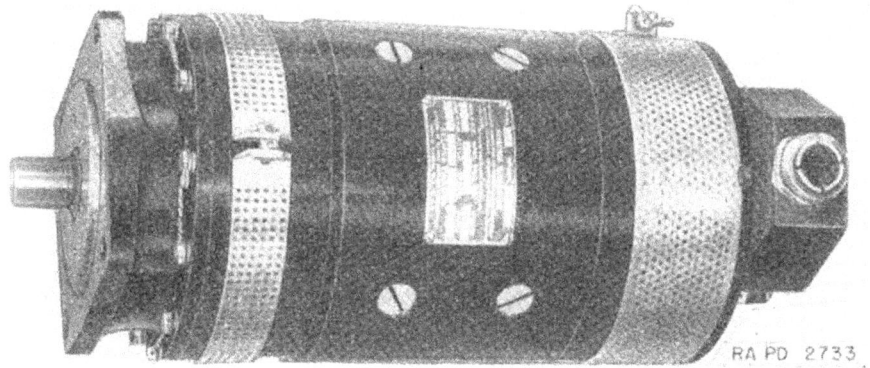

FIGURE 27.—Generator.

(3) Remove the locking wire or cotter keys and the nuts holding the starter to the accessory case.

(4) Remove the starter.

(5) To replace the starter reverse the above procedure.

e. Lubrication.—All starters are sufficiently lubricated at the time of issue and should not require any further lubrication between major engine overhaul periods.

53. Generator (fig. 27).—*a. Description.*—The generator is flange-mounted on the accessory case below the starter and is held in place by four bolts. The voltage control box described in paragraph 103 controls the current output of the generator.

b. Lubrication.—Generators are properly lubricated at engine overhaul periods and should not require any additional lubricant between overhaul periods.

c. *Replacement.*—A faulty generator should be replaced as follows:

(1) Remove the terminal guard and disconnect leads.

(2) All leads should be marked to identify their proper position for replacement.

(3) Remove the four nuts holding the generator to the accessory case.

(4) Loosen the generator retaining bracket nut and slide the generator to the rear.

d. *Oil leakage.*—If excessive oil is present in the generator at the time of inspection, the generator should be replaced.

54. Fuel pump (fig. 22).—*a. Description.*—The fuel pump is mounted on the engine accessory case to the right of the generator.

b. *Maintenance.*—At the 50-hour inspection, check for leakage at the joints and fittings.

c. *To remove and replace fuel pump.*—In the event the fuel pump fails to supply a sufficient quantity of fuel, the pump shoul be replaced with a serviceable unit.

d. The following procedure will be observed for removal and replacement of the fuel pump:

(1) Close fuel shut-off valves. The two main valves are located in bulkhead below oil tank.

(2) Remove cap screws and open rear doors to engine compartment.

(3) Disconnect fuel lines attached to pump.

(4) Remove locking wires and nuts holding the pump to the accessory case.

(5) Remove pump.

(6) To replace pump reverse the procedure.

55. Carburetor (fig. 22).—*a. Description.*—The carburetor is attached directly to the intake manifold in the bottom of the engine crankcase.

b. *Maintenance.*—Once the carburetor is properly installed, very little attention is needed between major engine overhauls. A fuel strainer is located at the rear center of the carburetor and may be reached by the removal of the large square-headed plug at the bottom of the carburetor. A small squareheaded plug is provided as a drain in the bottom of each float chamber. The strainer and drain plugs should be removed frequently to remove the accumulated dirt. The fuel supply should be shut off before removing the plugs. The entire carburetor should be inspected to see that all parts are tight and properly safety-wired.

c. *Replacement.*—Should it become necessary to remove or replace the carburetor the following procedure should be followed:

(1) Check to see that fuel shut-off valves are closed.
(2) Disconnect the fuel line and drain lines.
(3) Disconnect the fuel cut-off and accelerator linkage.
(4) Disconnect air intake connections at flexible connector.
(5) Remove the locking wire and the four bolts holding the carburetor to the intake manifold base and remove carburetor. Disconnect air intake pipes.
(6) To replace carburetor, reverse the procedure outlined above.

d. Adjustment.—After a new carburetor is installed it should be adjusted to the engine. Two adjustments are provided, one for idling speed and the other for mixture quality. The throttle adjustment on the carburetor (see fig. 22) is set so that the engine idles between 300 and 400 rpm as read on the tachometer. The mixture control lever is moved to the right (lean) until the engine begins to run unevenly. The lever is then moved one notch at a time to the left (rich) until the engine operates evenly.

56. Air cleaners (fig. 36).—*a. Description.*—Two air cleaners of the oil bath type are provided, one at the right and one at the left rear engine compartment.

b. Maintenance.—After each 50 hours of normal engine operation or daily in extremely dusty operation, service the cleaner in the following manner (see fig. 36):
(1) Remove the six wing nuts from bottom plate of cleaner, and take off bottom plate.
(2) Remove cup and flute assembly by lifting up slightly and twisting in a counterclockwise rotation. Take flute assembly out of cup.
(3) Empty dirt and oil from cup, and clean thoroughly. Refill cup to proper level with oil specified in figure 12. Care should be taken not to fill cups over the oil level mark.
(4) When replacing the cups, make sure that the cup assembly is properly latched.
(5) Place the bottom plate on the cleaner and fasten with thumbscrews.

c. Replacement.—In the event it becomes necessary to remove or replace the air cleaners, proceed as follows:
(1) Remove rear top plate over engine compartment.
(2) Disconnect rubber outlet connection to manifold (two hose clamps).
(3) Remove cups from cleaner.
(4) Remove cap screws connecting cleaner to sponson side.
(5) Lift out cleaner.

d. Clean filter.—Wash filter element with cleaning fluid or gasoline, dry, and blow out with compressed air in reverse direction to normal air travel.

57. Governor (fig. 22).—The governor is used to regulate and limit the speed of the engine after the engine has been manually accelerated. This prevents dangerous overspeeding. The unit is located on the accessory case below and slightly to the left of the generator. The governor is sealed and should not be tampered with.

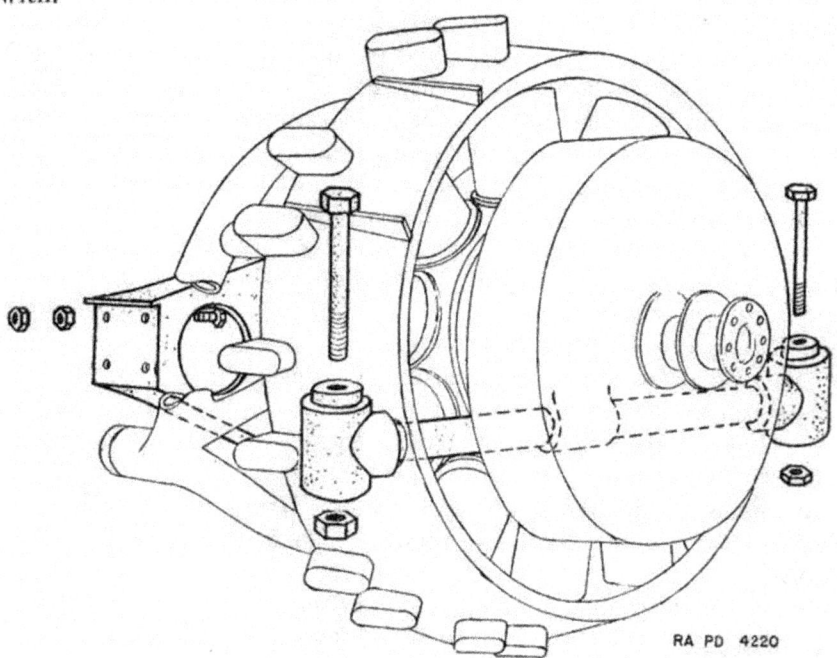

FIGURE 28.—Mounting of engine.

58. Engine lubrication and lubricants.—*a. Specifications.*—Refer to section IV, chapter 1.

b. Engine system (fig. 9).—Oil is drawn from the oil tank and delivered under pressure by the oil pump through a filter to all the necessary bearings and other parts of the engine through drilled passages. The engine scavenge pump returns the oil to the oil cooler. The pressure relief valve on the pump is adjusted to prevent the oil pressure from exceeding 80 pounds.

59. Replacement of engine.—*a.* Facilities must include the necessary engine stand and hoisting facilities.

b. To remove engine (see figs. 29 and 30).—(1) Open master switches beneath turret.

MEDIUM TANKS M3, M3A1, AND M3A2

TM 9-750
59

(2) Shut off all gasoline shut-off valves inside and outside hull. Remove outside handles.

(3) Remove bolts holding armor plates over engine ($7/8$-inch wrench). Open engine doors.

(4) Remove grill from front edge of armor plates over engine ($9/16$-inch socket).

(5) Remove cap screws holding shrouding to front section of armor plate ($9/16$-inch wrench).

(6) Attach sling and remove armor plates with **A**-frame.

(7) If necessary, detach Homelite exhaust pipe from clamps.

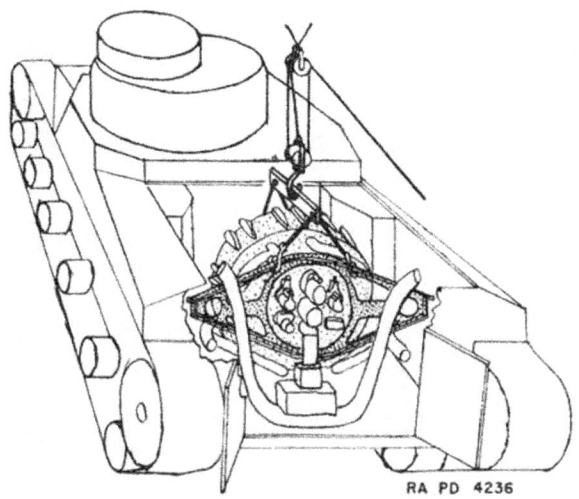

FIGURE 29.—Hoisting of engine, end view.

(8) Remove left air cleaner and connecting hoses.

NOTE.—Remove 6 wing nuts and bottom plate; remove 3 cups; loosen clamps and large intake hose; loosen clamp from engine breather hose and separate from intake duct; remove 10 cap screws holding air cleaner body to hull ($1/2$-inch wrench) and remove cleaner.

(9) Remove right air cleaner. (Same except that there is no breather from engine.)

(10) Disconnect gasoline "in" line to pump and return line at carburetor ($7/8$-inch open end wrench). Plug lines.

(11) Disconnect oil "in" and "out" lines ($1 3/8$-inch open end wrench) and plug lines.

(12) Disconnect Hycon lines at top and bottom of Hycon pump ($25/32$-inch). Plug opening with cloth.

(13) Disconnect oil temperature gage cable ($11/16$-inch). Plug openings.

71

TM 9-750

(14) Disconnect clip holding temperature gage cable at generator (⅜-inch). Replace screw.

(15) Disconnect tachometer drive cable by unscrewing knurled nut.

(16) Remove conduit and wires *at generator*. (Pliers, screw driver, and ⅜-inch open end wrench.)

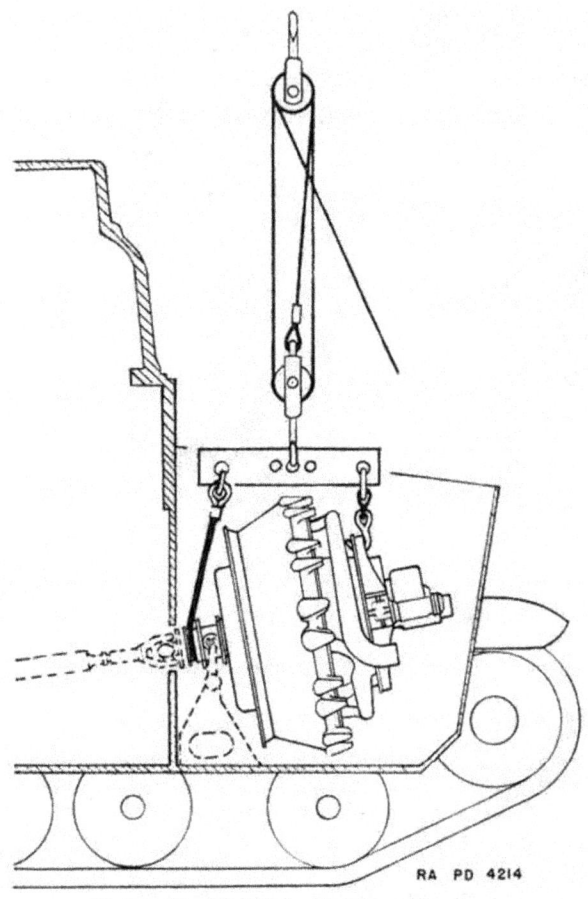

FIGURE 30.—Hoisting of engine, side view.

(17) Remove conduit and wire from starter *at starter*. (Pliers, offset screw driver and ⁹⁄₁₆-inch wrench).

(18) Remove magneto ground wires and booster wire at junction box.

(19) Remove upper shroud section (⁹⁄₁₆-inch wrenches).

(20) Remove inspection plate under hull (¾-inch wrench and hydraulic floor jack).

(21) Disconnect conduit and wire at cut-off solenoid and remove clips from banjo.

(22) Remove accelerator rod clevis pin beneath engine.

(23) Move slip sleeves on exhaust pipes away from joints ($7/16$-inch wrenches).

(24) Remove engine steady rest bolts ($1\frac{1}{4}$-inch socket and extension; $1\frac{1}{8}$-inch open end wrench).

(25) Disconnect ground strap from banjo ($9/16$-inch wrenches).

(26) Disconnect primer line at right of cut-off solenoid ($7/16$- and $3/4$-inch wrenches).

(27) Disconnect oil pressure line at right of cut-off solenoid ($5/8$- and $7/8$-inch wrenches).

(28) Remove starter solenoid and bracket from brace ($7/16$-inch wrenches).

(29) Disconnect breather line (crankcase to expansion tank) at forward end of metal tube located over top of engine. (Hose clamp connection.)

(30) Disconnect 2 muffler pipe support brackets from banjo ($9/16$-inch wrench).

(31) Remove fire extinguisher bracket, tube, and horn assembly from bulkhead ($7/16$-inch; 1-inch; $1\frac{1}{4}$-inch).

(32) Remove bolts from companion flange ($9/16$-inch wrenches).

(33) Install lifting sling on engine.

(34) Remove banjo bolts (pliers, two $9/16$-inch wrenches).

(35) Lift engine from compartment with **A**-frame. Set in stand. Insert banjo bolts and **C**-clamps, if necessary.

NOTE.—Keep all nuts, bolts, screws, etc., in tool box until ready for installation again.

c. To install engine.—Reverse the above procedure.

60. Valve adjustments (fig. 31).—Valve timing check and clearance adjustment (with engine cold):

a. Remove rocker box covers and gaskets from all cylinders.

b. Remove rear spark plugs.

c. Remove two adjacent $3/4$-inch hex elastic stop nuts which hold the flywheel and fan together, and attach the timing scale portion of the timer using the nuts just removed. Insert the top dead center indicator in the spark plug hole of the number one cylinder and determine the top dead center as follows:

(1) Turn the crankshaft in the normal direction of rotation (counterclockwise when looking at the flywheel or forward end of the engine) until the pointer of the top center indicator registers zero.

TM 9-750

ORDNANCE MAINTENANCE

(2) Clamp the timing pointer on the engine cowl in such position that the point is near the leading edge of the scale previously attached. Note the reading on the timing scale.

(3) Continue to turn the crankshaft in the same direction until the top center indicator has gone past and returned to the zero mark. Note the reading on the timing scale.

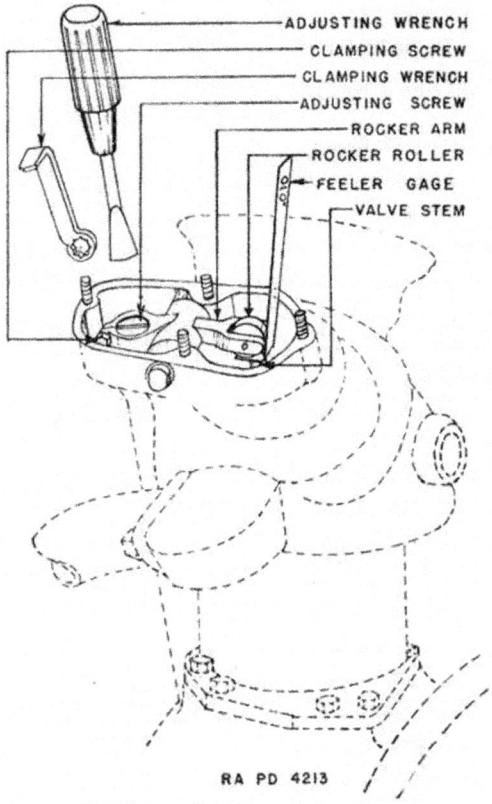

FIGURE 31.—Adjustment of valves.

(4) Turn the crankshaft backward about one-quarter of a revolution and then turn it forward until the timing pointer indicates a point on the timing scale midway between the two readings previously obtained. The No. 1 piston is now on the top dead center.

(5) Readjust the timing pointer on the cowl so that it indicates exactly zero on the timing scale.

d. Loosen the No. 9 cylinder intake valve clamping screw on rocker arm and turn the adjusting screw until the thread is flush with the rocker arm.

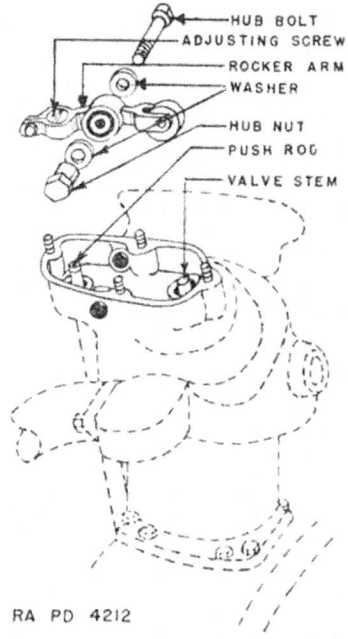

FIGURE 32.—Replacement of valve rocker arm.

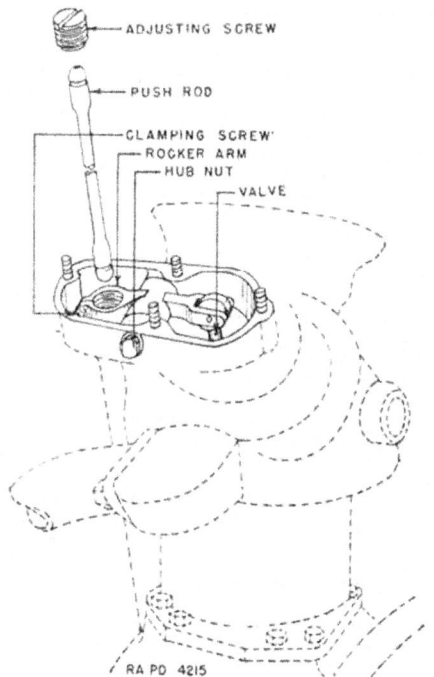

FIGURE 33.—Replacement of valve push rod.

e. Rotate the crankshaft in the direction of rotation until No. 9 cylinder intake valve just begins to open. Tap the push rod end of the rocker arms of No. 1 cylinder with a piece of fiber or a mallet. Set the clearance of both valves of No. 1 cylinder at 0.070 inch by loosening clamping screw and turning adjusting screws.

f. Clasp the rocker roller of No. 1 intake valve with fingers. Rotate the crankshaft in direction of rotation and note the point at which the roller binds and can no longer turn. This is the joint at which the intake valve opens and should occur from 3° to 7° before top center. Under the above conditions the exhaust valve will close from 2° before top center to 2° after top center.

g. Should valve timing not agree with that specified above, then a recheck should be made and if the timing is still found to be off, ordnance personnel should be advised to effect adjustment of the timing.

h. If the timing is correct, proceed to adjust the valve tappet clearance under the rocker arm rollers to 0.010 inch with the piston at top center at the beginning of the power stroke (fig. 31). Also do this for cylinder No. 9 which was disturbed as per *d* above.

i. Check all other cylinders for proper clearance and adjust to 0.010 inch if necessary.

j. Replace gaskets and rocker box covers on all cylinders. Replace spark plugs.

SECTION IV

FUEL SYSTEM

	Paragraph
Description	61
Shut-off valves	62
Primer	63
Fuel strainer	64
Gages	65
Gasoline system operation	66
Grades of engine gasoline	67

61. Description.—*a. Tanks.*—Four fuel tanks with a total capacity of 185 gallons are provided for the engine fuel system. Two vertical tanks, each of 32½-gallon capacity, are located in the front corners of the engine compartment, and two horizontal tanks of 60-gallon capacity are located one in each side of the engine compartment. A separate gasoline tank for the auxiliary generator with a capacity of 2½ gallons is located in the left rear corner of the crew compartment.

b. Drains.—The vertical tanks can be drained by removing a drain plug at bottom of the tank, directly beneath the filler cap. To

Figure 34.—Top view of engine compartment.

reach the drain plug, a 1½-inch pipe plug in the floor located beneath the drain plug must be removed. To drain the horizontal tank, remove the circular plate located at the rear and bottom of sponson, and remove drain plug.

c. Replacement.—Fuel tanks will be replaced when excess rust is present or when leaks develop. Fuel tanks may be removed as follows:

(1) *Vertical tanks.*—(*a*) Drain tanks.

(*b*) Remove front top plate.

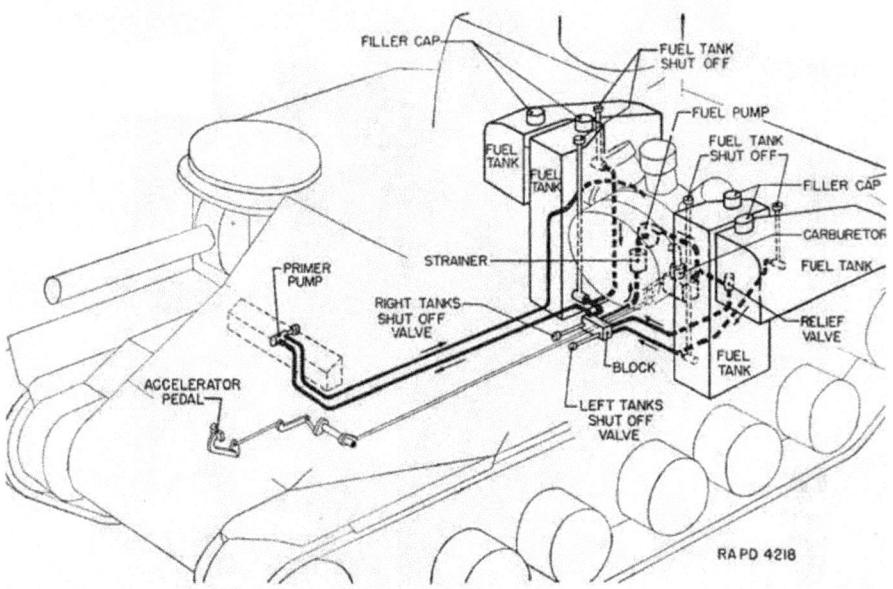

FIGURE 35.—Fuel system.

(*c*) Remove bottom plate at bottom of engine.
(*d*) Remove bulkhead brackets on each side.
(*e*) Disconnect fuel line connections.
(*f*) Screw out shut-off valve.
(*g*) Lift out tank.

(2) *Horizontal tanks.*—(*a*) Drain tanks.
(*b*) Disconnect gas lines and screw out shut-off valve.
(*c*) Remove engine.
(*d*) Remove side plate of sponson.
(*e*) Remove horizontal tank with top half of housing.

62. Shut-off valves.—*a.* Two valves (fig. 35) located on the rear wall of the fighting compartment under the engine oil tank and to the right of the power shaft tunnel control the flow from the right and

MEDIUM TANKS M3, M3A1, AND M3A2

TM 9-750
62-63

left fuel tanks. In addition, each tank is provided with a separate shut-off valve (fig. 35). The handles for the vertical tanks extend through the grill, near the turret, in the roof of the hull, while the handles for the horizontal tanks are placed in the countersunk holes in the roof of the hull, back of the filler caps.

b. The control valves under the oil tank in the rear of the fighting compartment should be turned off whenever the vehicle is stopped for any extended period of time.

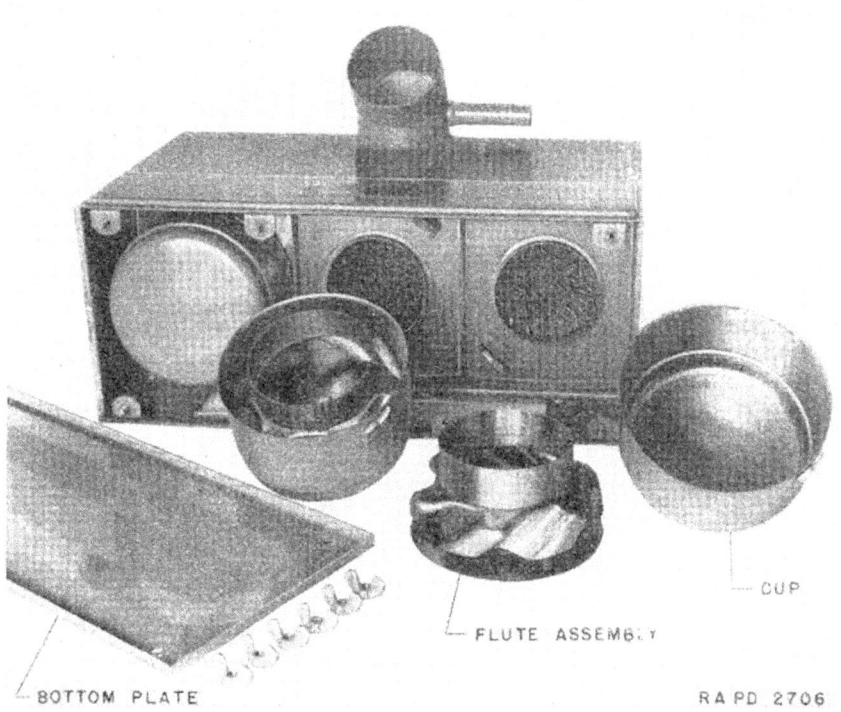

Figure 36. Vortox air cleaner.

c. The fuel can also be cut off electrically at the carburetor when the engine is idling, by means of a toggle type switch (fig. 8) located on the upper left-hand side of the instrument panel. Pressing the switch to the left automatically stops the flow of gasoline to the idling jet, stopping the engine.

63. Primer.—*a. Description and operation.*—A priming pump, located on the instrument board, provides a means of injecting a spray of fuel into the engine intake manifold to facilitate starting. The pulling stroke of the primer draws a charge of gasoline into the primer cylinder. During the return or charging stroke, the charge is delivered to a small distribution housing from which fine lines run to the top five intake manifold pipes.

  b. Maintenance.—If more than a few strokes are required to prime the engine, the leather packing on the instrument panel end of the plunger should be checked for leakage. To stop leakage, compress packing by half turns on the packing nut located behind the priming plunger button on instrument panel until the leakage stops.

  c. Primer pump removal.—(1) Disconnect inlet and discharge connection at primer.

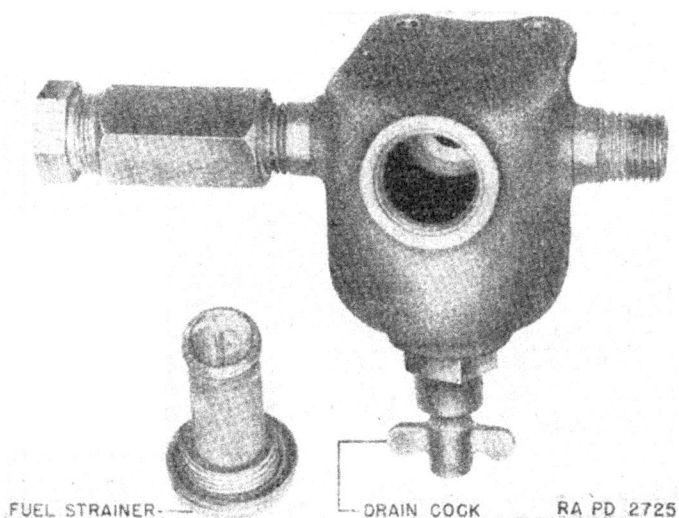

FIGURE 37.—Gasoline fuel strainer.

  (2) Loosen packing nut and remove primer plunger.
  (3) Unscrew large retaining flange type nut on front of instrument panel.
  (4) Remove from panel.

64. Fuel strainer.—*a. Description* (fig. 37).—The fuel strainer is of the sediment bowl type installed in the fuel system between the fuel supply valves and the fuel pump. It is located in the right rear corner of the engine compartment.

  b. Maintenance.—The strainer screen can be removed and cleaned by removing the $\frac{1}{2}$-inch square plug on the side of the strainer and can be drained by the needle drain valve at the bottom of the strainer.

65. Gages (fig. 38).—Two electrically operated fuel gages are mounted in a small box located near the oil cooler in the rear of the

TM 9-750
MEDIUM TANKS M3, M3A1, AND M3A2   65-68

main fighting compartment. The rear gage is connected to the large left tank and the front gage is connected to the large right tank. For wiring see figure 70. A toggle switch is provided on the front panel of the box to operate a small light for lighting up the gages. The gages are put into operation when the ignition switch is turned "on."

66. Gasoline system operation (fig. 35).—Gasoline from any tank or combination of tanks flows to a central header. From the header it is drawn through a strainer by a pump to the carburetor. Excess gasoline is bypassed through a relief valve and returned to

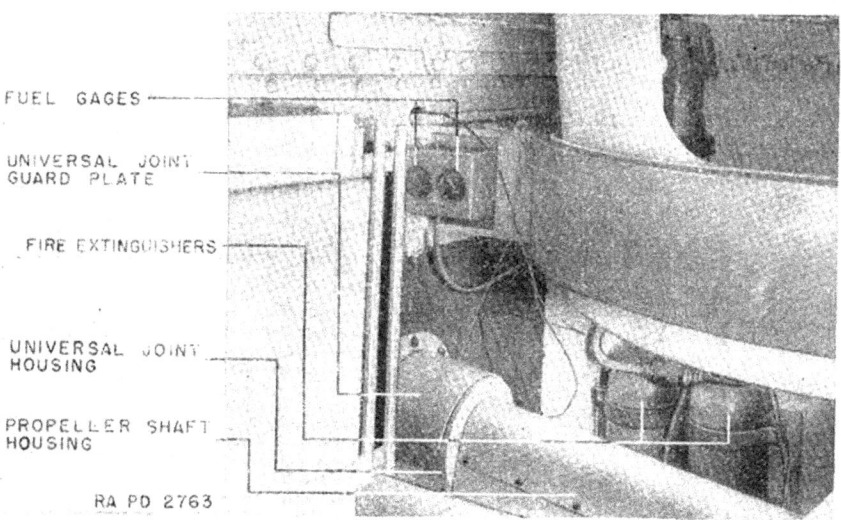

FIGURE 38.—Interior view, right side of bulkhead.

the header to prevent vapor lock and excess gasoline pressure. An additional gas line leads from the header to the priming pump located on the instrument panel.

67. Grades of engine gasoline.—Commercial motor gasoline having an octane rating of 91 is preferred and should be used. In emergencies, other gasoline with an octane rating of not less than 82 may be used.

Section V
COOLING SYSTEM

Paragraph
Description _____ 68
Inspection and maintenance _____ 69

68. Description.—The engine is cooled by an air blast produced by a fan mounted on the engine flywheel. The fan draws air through a grill in the roof of the hull and forces it between and around the finned cylinders of the engine. The warm air passes out through a screen above the engine doors on the rear of the tank. Air ducts are formed on the engine by baffles bolted around and between each cylinder and cylinder head. A shroud forms a further duct on the inlet of air through the grill.

69. Inspection and maintenance.—At the 100-hour inspection the cylinder heads and cooling fins on the engine cylinders and the surfaces of the shrouding and baffles will be cleaned thoroughly of all accumulated oil and dirt. All bolts on the fan ring, baffles, and fan cowling will be tightened.

Section VI

CLUTCH

	Paragraph
General	70
Operation	71
Care	72
Inspection in vehicle	73
Clutch pedal adjustment	74
Clutch release	75
To remove clutch	76
Inspection after disassembly	77
Clutch outer hub bearing	78
Clutch inner hub bearing	79
Clutch plates	80
Reassembling clutch	81

3. General.—The clutch serves as a means of applying engine power to the power train of the vehicle, which consists of propeller shaft, transmission, differential, final drives, and sprockets. The clutch is built into the engine flywheel and is located in the front of the engine.

71. Operation.—*a.* The clutch allows the engine to pick up load gradually after shifting gears. When the clutch is engaged, the springs inside the clutch housing exert their full pressure against the clutch plates, enabling the engine power to go through to the propeller shaft.

b. When the clutch pedal is depressed, the pressure plate is pulled back from the clutch plates, compressing the clutch springs, allowing the clutch plates to rotate free of the propeller shaft. Engine power is thus prevented from going through to the propeller shaft.

72. Care.—The clutch outer hub bearing (fig. 39) is greased as prescribed in figure 12. The clutch is disassembled and cleaned as shown in figure 41 at each 100-hour inspection.

MEDIUM TANKS M3, M3A1, AND M3A2

TM 9-750
72

Figure 39.—Engine fan and clutch.

73. Inspection in vehicle.—Faulty operation, such as slipping, grabbing, or rattling, as noted by the driver, should be reported to proper authority.

74. Clutch pedal adjustment.—*a.* As the clutch facing wears, the amount of pedal free play is reduced and in time this will result in the clutch pedal striking the floor plate and cause clutch slippage. It is, therefore, necessary to check the pedal clearance at intervals of 100 hours to provide sufficient free pedal play to permit engagement of clutch.

b. Correct adjustment of the clutch pedal is made by lengthening the adjusting link (fig. 40) so that ⅛-inch clearance exists between the clutch release bearing and flange when the clutch is engaged.

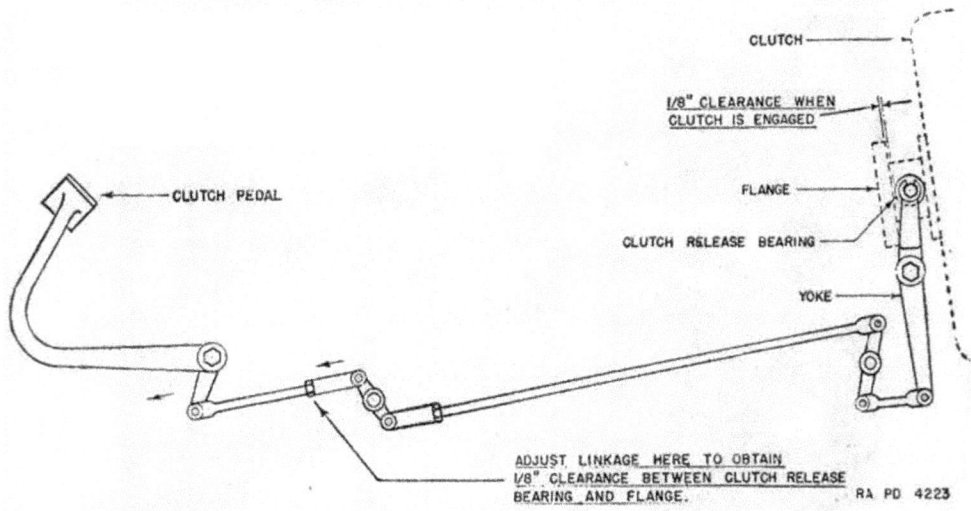

FIGURE 40.—Adjustment of clutch.

75. Clutch release.—*a.* The clutch is provided with a release device which holds the clutch in the "out" position (disengaged). The foot pedal and levers shown in figure 40 are part of the release mechanism. Release bearings should always be inspected when the engine is removed.

b. Clutch release bearings are lubricated at least once each day with a few drops of engine oil as prescribed in figure 12.

c. When release bearings are to be removed, first remove lubrication fitting and *remove screw which locks nut onto shaft*.

76. To remove clutch.—To remove the clutch for inspection and possible replacement of plates, bearings, or grease retainers, first remove the engine and clutch from the vehicle as directed in section

MEDIUM TANKS M3, M3A1, AND M3A2

TM 9-750
76-77

III and proceed as follows (see fig. 41). Punch mark pressure plate assembly before taking apart, as each unit is carefully balanced.

a. Remove flange nut (1).

b. Remove flange (2).

c. Remove nuts and cap screws that hold clutch housing assembly to flywheel (5). This includes six cap screws holding release arms and shim pads.

d. Remove clutch housing assembly (6).

e. After removing two screws holding plates from sliding off spindle, remove spindle (8).

f. Lift out the three metal plates and the two faced plates (9 to 13 incl.).

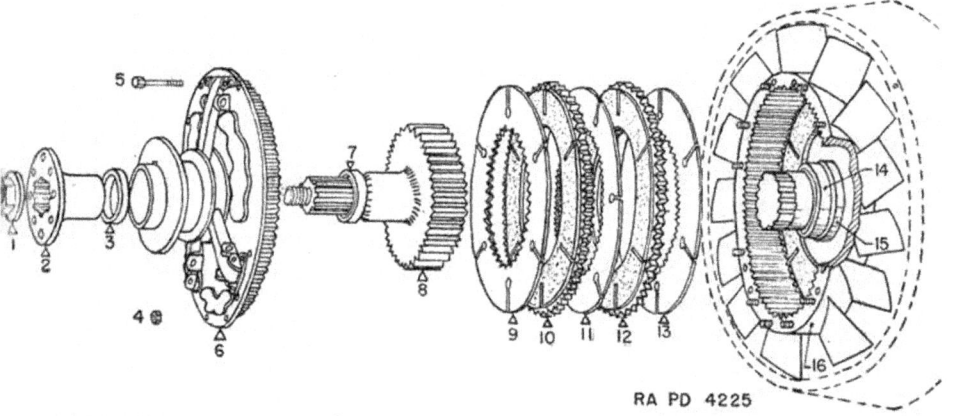

1. Flange nut.	7. Bearing, outer hub.
2. Flange.	8. Spindle.
3. Grease retainer.	9. Metal plate.
4. Housing nut.	10. Faced plate.
5. Housing cap screw.	11. Metal plate.
6. Housing.	12. Faced plate.
	13. Metal plate.
	14. Bearing, hub.
	15. Grease retainer.
	16. Flywheel.

FIGURE 41.—Disassembly of clutch.

g. Do not remove the flywheel from the engine crankshaft. Any wear of facing in flywheel will be reported to the ordnance maintenance personnel.

h. Disassembly of the clutch housing is not to be attempted.

77. Inspection after disassembly.—*a.* Wash all parts.

b. Inspect pressure plate and flywheel for any indication of scores on contact surface. Then check metal plates for burs and cracks.

c. Inspect faced clutch plates for worn or loose facing and loose rivets. Examine splines to see that they move freely within each

85

other. If splines are worn, notify ordnance maintenance personnel.

d. Maintenance diagnosis and remedial measures are as follows:

Symptom and probable cause	Probable remedy or action
(1) *Slipping.*	
(*a*) Weak spring action	Notify ordnance personnel.
(*b*) Facing torn loose from plate	Replace plate.
(*c*) Sticking pressure plate	Clean outer edge of pressure plate assembly.
(*d*) Excessive oil soaking of plates	Clean plates.
(2) *Grabbing.*	
(*a*) Oil on lining	Clean plates.
(*b*) Worn splines on spindle	Replace spindle.
(3) *Vibration.*	
Release yoke loose on stud	Replace yoke.

78. Clutch outer hub bearing.—*a.* Clutch outer hub bearing is shown in figure 41. Care should be used in removing this bearing to prevent damage. After removal, it should be cleaned in dry-cleaning solvent, inspected, packed with lubricant, or replaced with a new bearing if necessary.

b. When installing a hub bearing, it is essential that it be pressed in place in exact alinement with the seat in the clutch housing.

79. Clutch inner hub bearing.—*a.* The clutch inner hub bearing is shown in figure 41. The same precautions and treatment are to be followed in removing and replacing this bearing as for the clutch outer hub bearing.

b. The clutch inner hub bearing is lubricated only by packing at time of reassembly. Consequently, it is necessary that this bearing be carefully packed as prescribed in figure 12 before it is pressed back into place. An excessive amount of lubricant in the inner clutch hub bearing will prevent proper assembly.

80. Clutch plates.—Clutch plates provide the proper amount of slipping for gradual and smooth application of power when starting, and positive nonslipping drive when fully engaged. Worn or damaged clutch plates will be replaced with new or rebuilt plates.

81. Reassembling clutch.—The order of reassembly is to follow in reverse the steps used in taking the clutch apart, except that the spindle ((8), fig. 41) is replaced before the clutch plates (9 to 13 inclusive). Remember to replace screws and wire that hold plates on spindle. Care must be taken in replacing flange, as the lips of seal in sliding sleeve face out. Use shim stock or feeler gages to ease companion flange onto spindle so as not to rupture seal.

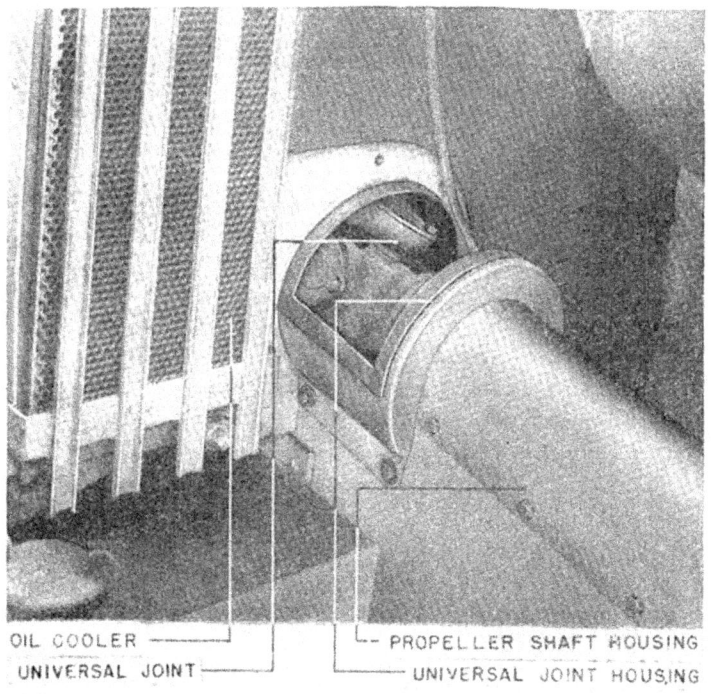
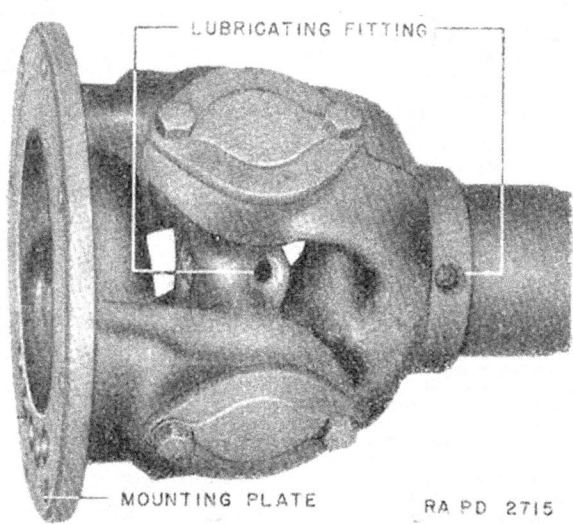

Figure 42. — Propeller shaft and universal joint.

Section VII

PROPELLER SHAFT

	Paragraph
General	82
Lubrication	83
Inspection	84

82. General.—The propeller shaft (fig. 42) transmits power from the clutch to the transmission and the rest of the power train. It is equipped with universal joints at either end to permit operating at an angle with the transmission and clutch. It is inclosed in a housing or "tunnel" in the center of the floor of the crew compartment.

83. Lubrication.—The universal joints and spline are equipped with small plugs which are replaced with pressure fittings for lubrication. Lubrication will be in accordance with figure 12. In order to reach these plugs, guard plates on the ends of the tunnel must first be removed by taking out the four cap screws holding each guard plate.

84. Inspection.—Universal joint bearings and the slip joint grease seals are the parts which are most apt to wear out. Excessive vibration, produced by the propeller shaft while it is in motion, should be reported to ordnance personnel. The slip joint grease seals should be inspected for leakage during the 100-hour check.

NOTE.—The using arms are authorized to remove and reinstall a propeller shaft, obtaining necessary instructions from ordnance personnel. The replacement of one propeller shaft with another propeller shaft must not be done by using arms, however, unless authorization is received from ordnance personnel. When installing a propeller shaft it is important that the marks on the slip joint line up.

Section VIII

TRANSMISSION, DIFFERENTIAL, AND STEERING BRAKES

	Paragraph
Transmission and differential	85
Steering brakes	86
Adjustment of steering brakes	87
Adjustment of parking brake	88
Replacement of gear shift lever	89
Lubrication	90

85. Transmission and differential.—*a. Transmission.*—The transmission has five forward speeds and one reverse speed. It is synchromesh in second, third, and fifth gears, and constant mesh in first and reverse. A parking brake is built into the transmission and is operated by a lever to the left of and behind the driver. This brake will be used only after the vehicle has been brought to a stop.

b. Differential.—The differential is called a "controlled differential" not only because it serves to transmit engine power to the final drive units, but also because it is controlled by the steering brakes, in order that the tank might be steered.

86. Steering brakes.—The steering brake assemblies are located on either side of the differential, between it and the final drives. Older models have a hydraulic booster (Hycon) system attached to steering levers.

87. Adjustment of steering brakes.—The adjustment of the steering brakes is extremely important and only specially qualified mechanics should be permitted to perform it. Figure 45 illustrates the adjustment of the steering brake. Both bands should be adjusted equally and the linkage to the levers then set so that the bands contact the drum at the same time when the steering levers are pulled back evenly.

a. Remove the plug and gasket from the brake drum housing cover.

b. Insert through the hole a $1\frac{1}{16}$-inch deep socket wrench (supplied in the tank mechanic's set) and engage this wrench over the brake adjusting nut.

c. Turn the brake adjusting nut one-half revolution clockwise to tighten the brake band.

NOTE.—The brake adjusting nut has a cylindrical surface on the pressure end instead of the usual flat face. It is, therefore, imperative that this cylindrical surface be seated firmly against the cross pin when the above adjustment is completed.

d. Check adjustment by pulling back on the steering lever which is connected to the brake band.

e. In older models, the hydraulic booster system greatly influences the brake operation. It may be found that the above adjustment does not produce satisfactory operation. In this event ordnance maintenance personnel should be notified.

88. Adjustment of parking brake (fig. 43).—The parking brake depends upon a toggle joint for its locking action. This joint consists of an eccentric on the end of the cross shaft to which is fastened an eye and a link. The adjustment of this linkage should be such that application of the brake lever will cause the toggle action to go past center and into the locked position after seating the brake shoe lining firmly against the conical brake drum. To adjust the parking brake proceed as follows:

TM 9-750

ORDNANCE MAINTENANCE

a. Remove the cotter pin from the eccentric end of the lever cross shaft and remove the washer.

b. Pull out the cross shaft until the eccentric end of the shaft passes out of and clears the eye which is threaded into the link.

c. Remove the spacer from the shaft.

d. Turn the eye counterclockwise, thus screwing it out from the link the desired amount. For normal adjustments to take up for

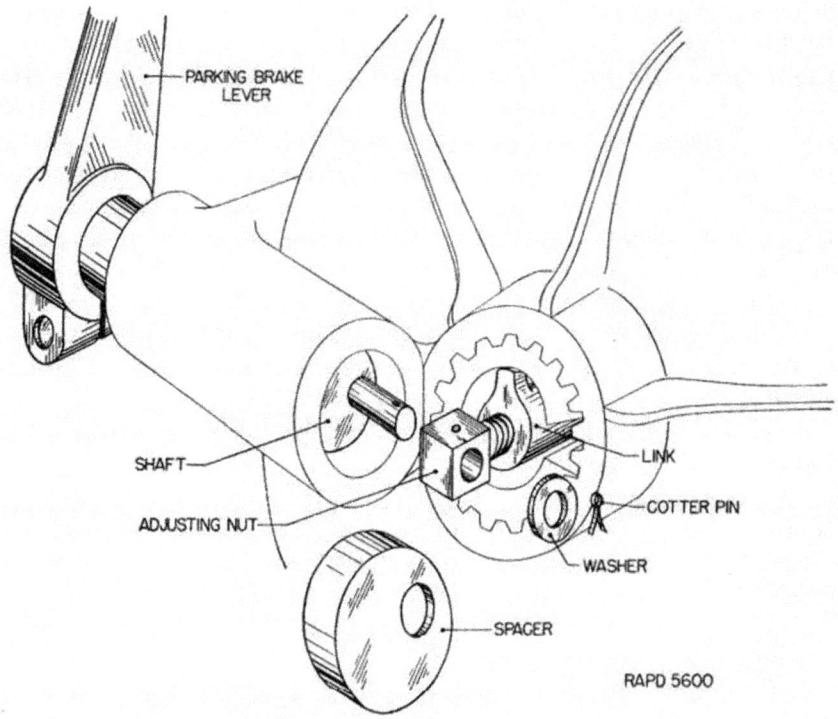

FIGURE 43.—Parking brake adjustment.

wear, it is best to turn the eye one-half turn at a time, slide the cross shaft back into position, try the brake, and repeat this until the condition described above is obtained.

89. Replacement of gear shift lever (see fig. 46).—If the gear shift lever becomes broken or bent, it can be replaced as follows:

a. Remove cotter key from nut.

b. Remove nut.

c. Drive out bolt.

d. Lift out lever assembly.

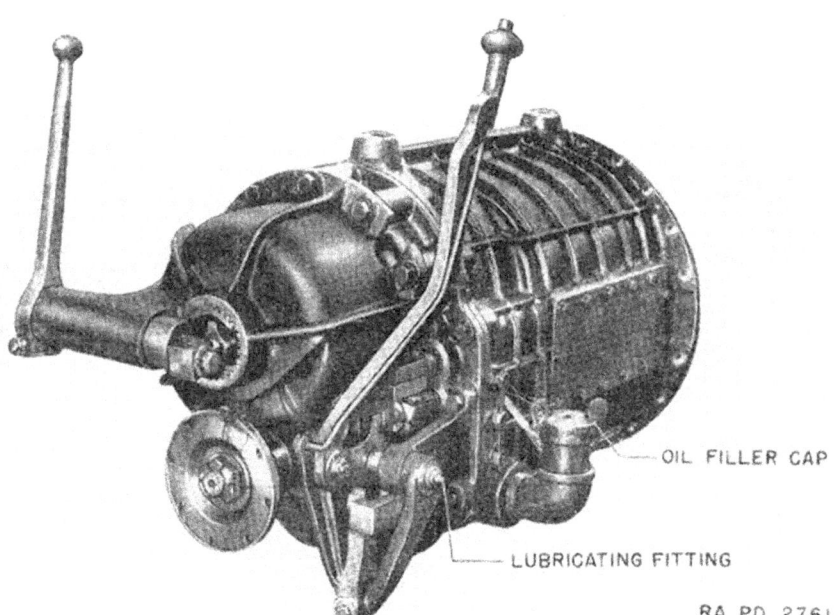

Figure 44.—Transmission.

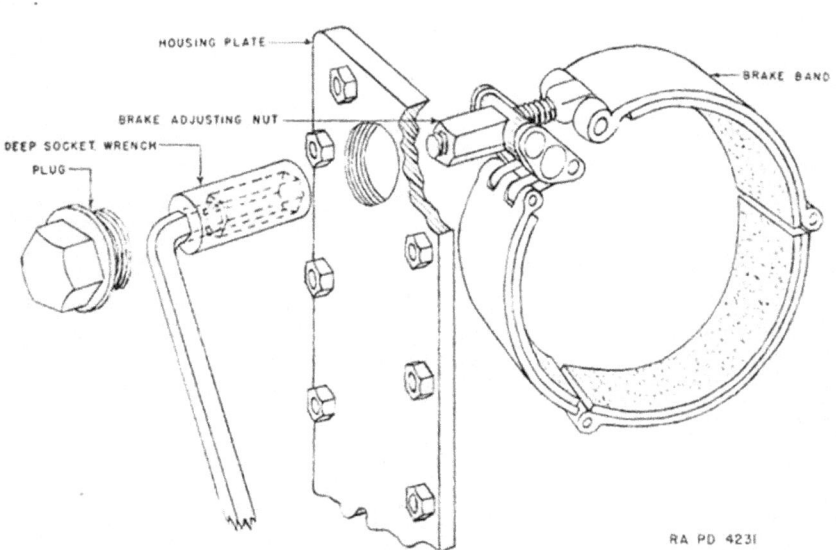

Figure 45.—Adjustment of steering brake.

TM 9-750
89-90   ORDNANCE MAINTENANCE

e. Replace in reverse order. One arm of the yoke is straighter than the other. It is absolutely necessary that the straightest arm be on the left side of the yoke, when replaced, in order that the shifting lever selecting finger be able to contact all shifter rails.

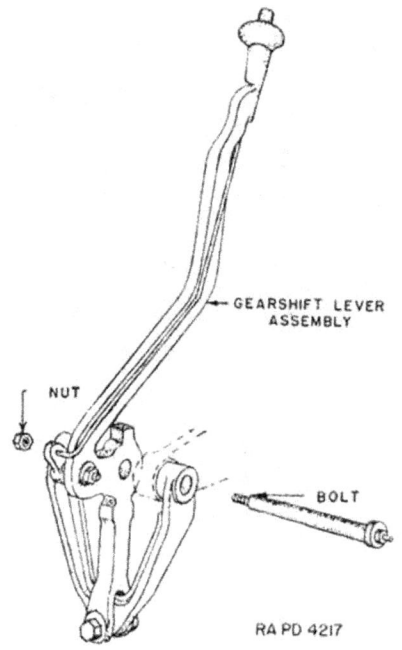

Figure 46.—Replacement of gear shift lever.

90. Lubrication.—*a. Transmission.*—(1) The transmission contains 26 quarts of oil, which is circulated by a small pump built into the transmission case. Oil is added and the level is checked by the removal of the filler cap, which is located to the right and to the rear of the driver's seat. A bayonet type oil gage is built into the cap. Oil is hot and foaming at the end of a run. It is necessary to let the tank stand before the level is checked and additions are made to keep the level up to the full mark. Use oil as specified in figure 12.

(2) The transmission oil will be drained at intervals, as indicated in section IV, chapter 1, by the removal of a plug in the bottom of the case. This plug is reached from underneath the tank, through an opening in the floor directly below the drain plug.

92

b. Differential.—(1) The differential and the final drives are lubricated by a single system. The oil level is checked by means of the plugs on the outside of the tank. Both plugs must be checked, unless the tank is on absolutely level ground. The oil level should be up to one-half inch below the bottom of the plug. The level of the oil will be checked at the end of each day's run. Oil of the proper grade (fig. 12) will be added to keep the level up to the full mark. Oil may be added at either the filler plug inside the tank or by removing the plugs in the front of the tank, on the outside in the final drive housings.

(2) The oil in the differential and final drives will be drained at intervals indicated in section IV, chapter 1. Two drain plugs are located on the outside of the tank in the bottom of each final drive housing. Another plug is located under the differential. All plugs must be removed. One hundred twenty quarts of oil of the grade shown in figure 12 will be required for refilling the differential and final drive.

SECTION IX

FINAL DRIVE

	Paragraph
Description	91
Lubrication	92
Removal of final drive unit	93

91. Description.—The final drive transmits the power from the controlled differentials to the hub of the driving sprocket through a set of reduction gears. The final drive and brake housings are bolted to each side of the differential case, and the driving sprockets are bolted to the hub. Both final drives are interchangeable, left for right, including cover, gasket, gears, and shaft.

92. Lubrication.—The final drive lubrication system is a part of the differential system and all details will be found in section VIII.

93. Removal of final drive unit (see fig. 47).—*a.* Slacken and disconnect the track under the drive sprockets.

b. Drain oil from the final drive and differential.

c. Remove the hub assembly by taking off the eight retaining nuts.

d. Remove the 22 studs which secure the final drive unit to the differential housing.

e. Withdraw the final drive unit.

f. To install unit reverse the procedure. It is advisable to replace gasket. Install final drive shaft with tapped end out.

TM 9-750
93-95   ORDNANCE MAINTENANCE

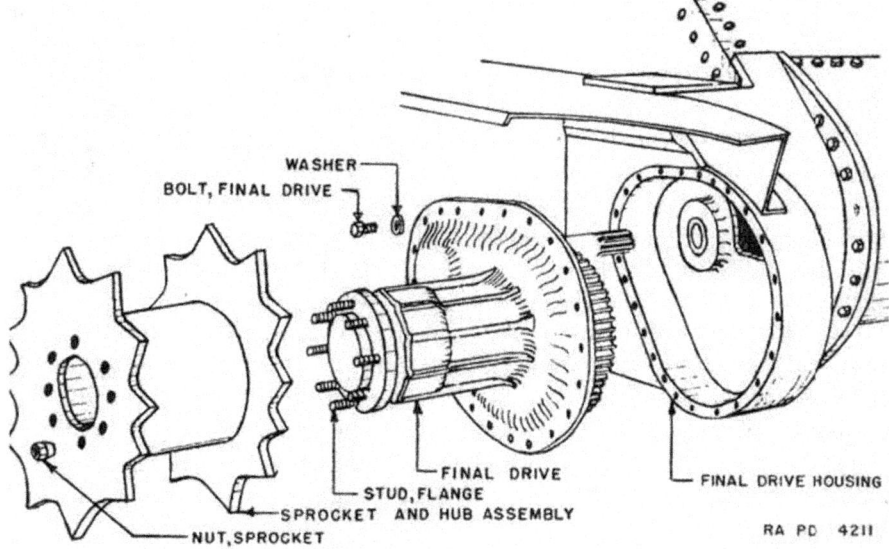

FIGURE 47.—Final drive and brake housing.

SECTION X

TRACKS AND SUSPENSIONS

	Paragraph
Description and operation	9
Drive sprockets	9
Bogies	9
Tracks	9
Idlers	9
Grousers	9

94. Description and operation (fig. 48).—Six two-wheeled rubber-tired bogies or suspensions, bolted to the hull, support the vehicle on springs. The tracks are driven by sprockets on the front of the tank. Two adjustable idlers at the rear end of the hull are provided to maintain constant tension on the tracks. The weight of the upper portion of the track is carried by steel track-supporting rollers on the top of the suspension bracket.

95. Drive sprockets (fig. 47).—*a. Description.*—Hub assemblies are formed by bolting two sprockets to a hub, which in turn is bolted to the flange end of the final drive shaft. These sprockets and hub assemblies are interchangeable as units, and should be transposed between the right and left final drive shafts when the teeth have been appreciably worn.

b. To replace the hub assembly (see fig. 47).—(1) Slacken and disconnect the track below the sprocket.

94

(2) Remove the eight hub retaining nuts.

(3) Remove the hub assembly.

(4) To replace the hub assembly, reverse procedure. Connect the track and adjust tension.

96. Bogies (fig. 48).—*a. Description and operation.*—The bogies are the supporting and conveying units and are sometimes called trucks or suspensions. The movement is transferred from the wheels to the arms and levers and is absorbed by springs. On top of each bogie assembly is a single steel roller to support and carry the upper portion of the track. The roller assemblies are bolted in seats on top of the bogie brackets.

b. Lubrication.—Lubrication of the wheels and the track supporting rollers is through lubrication fittings. Relief valves are provided to prevent injury to oil seals. See section IV, chapter 1, for lubrication instructions.

c. Care.—Pieces of gravel or stone have a tendency to wedge between the bogie tire and the bolt that connects the lever arms, resulting in rapid deterioration of the tires by the formation of deep grooves. Frequent examination is required to prevent this damage.

d. To replace bogie wheels (fig. 49).—Bogie wheels can be replaced without disconnecting or loosening the track.

(1) Place bogie wheel lift under bogie bracket as shown in figure 49. Drive the tank upon bogie wheel lift. Caution must be exercised that the tank is moved slowly and stopped in the proper position.

(2) Remove the cotter pin and nut on the back end of the bogie wheel gudgeon (fig. 50).

(3) Remove gudgeon with slide-hammer gudgeon-puller. If it is too tight, use the screw type gudgeon-puller. Care should be taken not to damage the Woodruff key, used to prevent the spindle from turning, or the thread on the end of the spindle.

(4) Install in reverse order; when replacing bogie spindle or gudgeon be sure that it is well oiled.

e. Replacing wheel bearings and oil seals.—(1) Remove spacers with oil seals and drive out bearings and spacer with copper drift.

(2) To replace bearings and oil seals—

(*a*) Place bearing in one side of wheel.

(*b*) Place spacer in wheel.

(*c*) Push other bearing in wheel until it rests against spacer.

(*d*) With two oil seals in place on each spacer leather portion nearest the center line of the wheel, place one spacer in each side of wheel.

Caution.—Make sure oil seals are on spacer correctly.

Figure 48.—Bogie wheels and supporting roller.

MEDIUM TANKS M3, M3A1, AND M3A2

TM 9-750
96

f. To replace volute springs.—(1) Slacken and disconnect the track.

(2) Remove roller from top of suspension.

(3) Remove plugs above each spring and insert jack studs into threaded plate at bottom of spring.

(4) Relieve tension on wheels by screwing jack nut on the stud against the suspension bracket as shown in figures 53 and 54.

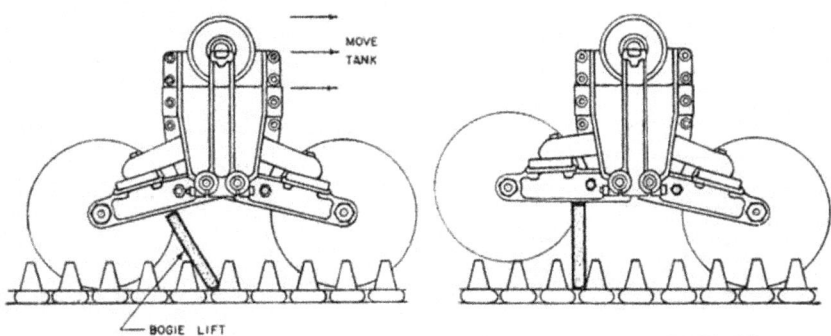

FIGURE 49.—Use of bogie left to raise wheel for removal.

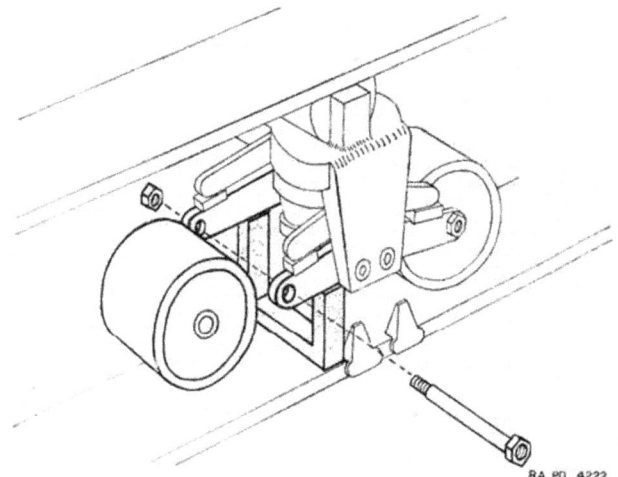

FIGURE 50.—Removal of bogie wheel.

(5) Remove gudgeon clamp screws.

(6) Raise tank off wheels, placing jack at edge of hull floor.

(7) Remove bogie arm gudgeons, using a slide hammer or screw type puller. Wheels and arms are now free and can be removed.

(8) Release springs by unscrewing jack nut.

NOTE.—When unscrewing jack nut to release spring, the socket wrench handle on the jackscrew should be held to prevent the jackscrew from turning

TM 9-750

ORDNANCE MAINTENANCE

out of the spring plate. As an extra precaution, a jack placed under the spring plate can be used together with the jack nut when releasing the tension of the spring.

(9) When tension is released, remove the jackscrew and the spring will fall out.

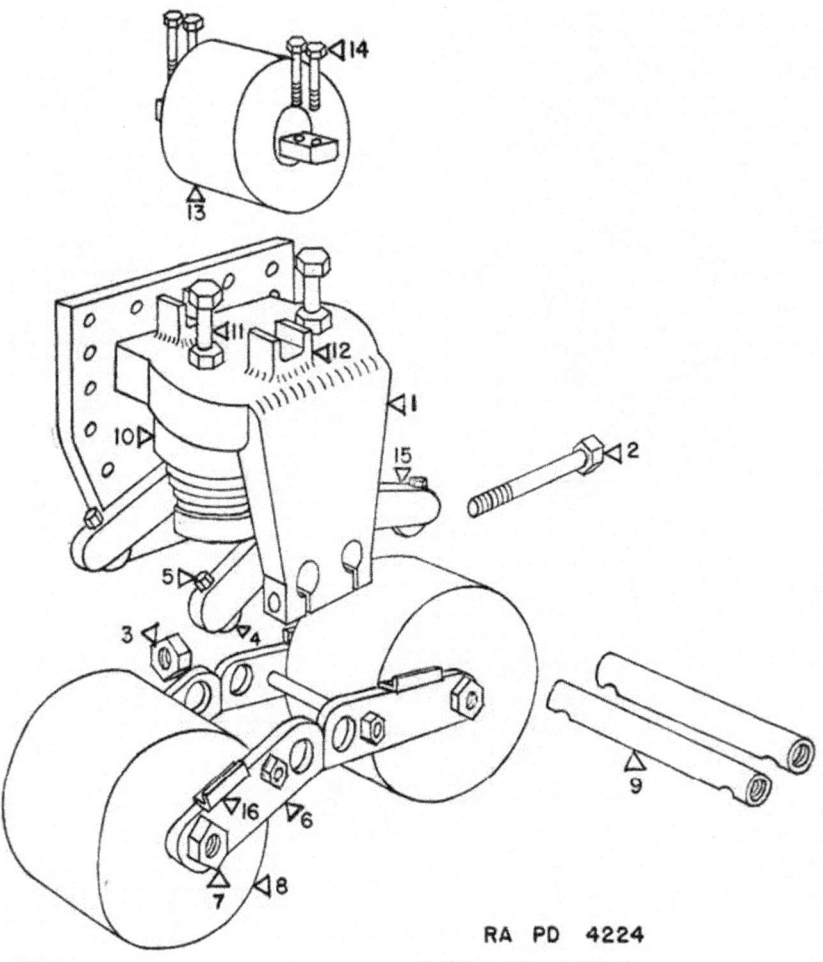

RA PD 4224

1. Frame.
2. Bolt.
3. Nut.
4. Wearing plate.
5. Cap screw, wearing plate retainer.
6. Lever arm.
7. Wheel gudgeon.
8. Wheel.
9. Suspension gudgeon.
10. Volute spring.
11. Spring compression bolt.
12. Roller support.
13. Track-supporting roller.
14. Roller axle bolt.
15. Spring retainer assembly.
16. Wearing plate.

FIGURE 51.—Bogie components.

MEDIUM TANKS M3, M3A1, AND M3A2

TM 9-750
96-97

(10) To replace the spring, reverse the procedure, using jack to raise lever spring plate, and *spring* it in place.

g. Wearing plates.—Replace worn or damaged wearing plates as shown in figure 52.

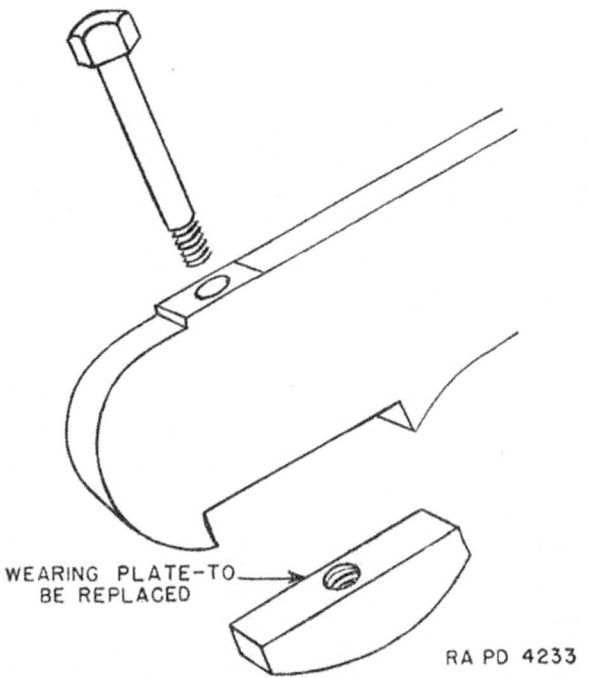

WEARING PLATE-TO BE REPLACED

RA PD 4233

FIGURE 52.—Removal of wearing plate from bogie lever arm.

97. Tracks.—*a. Description* (fig. 55).—Most of the medium tank tracks are composed of rubber track shoe units, each assembled with two pins. The units are assembled into an endless track by the steel end connections which are secured to the ends of the pins by wedges. The steel end connections also serve as guides to keep the track in alinement with the bogie wheels, idlers, and drive sprocket. The outer ends of the connections serve as driving lugs and engage the teeth of the final drive sprocket. Some of the tanks are provided with rubber jointed tracks having a steel wearing surface of interrupted grouser design and an inside cushioning surface of rubber.

b. Maintenance.—The rubber track shoes are of the reversible type and should be reversed when worn to such an extent that further wear will cause the tubular section of the metal link to become

TM 9-750
ORDNANCE MAINTENANCE

dented or deformed. Damaged or unserviceable units should be removed from the track and replaced with serviceable ones.

c. Removal of track (fig. 55).—(1) Make sure that tank is on level ground.

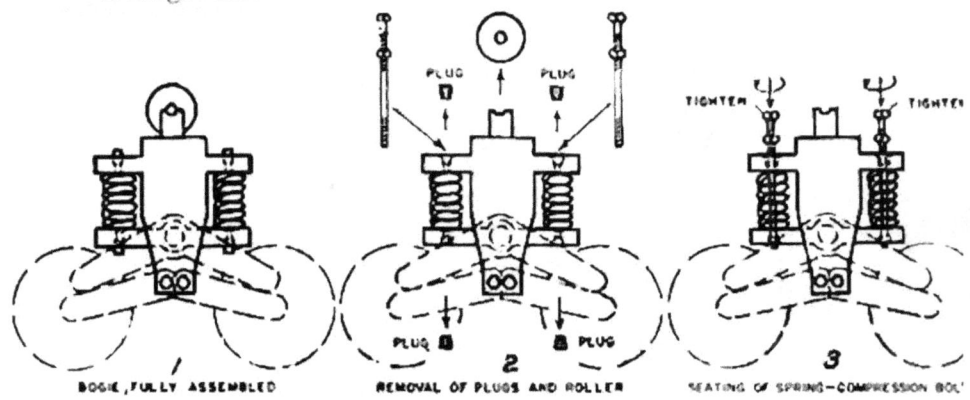

FIGURE 53.—Disassembly of bogie: first steps.

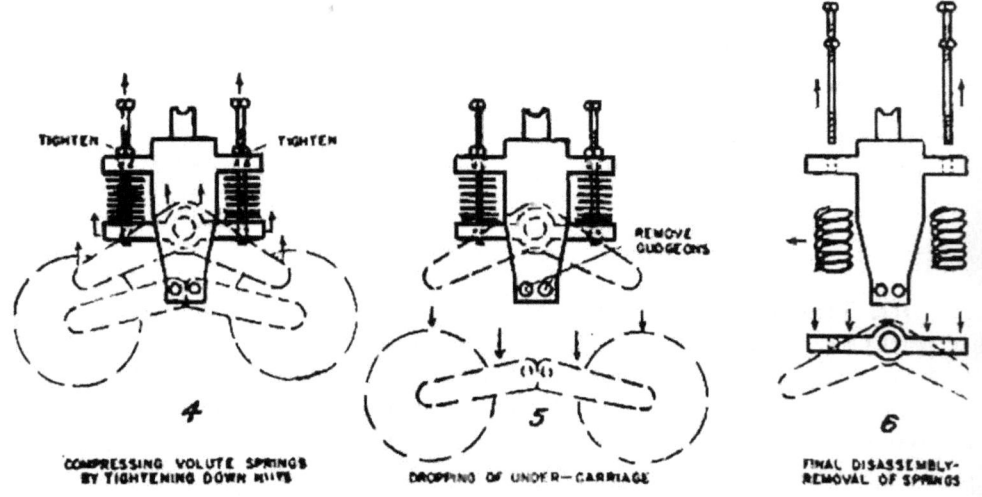

FIGURE 54.—Disassembly of bogie; final steps.

(2) Release the track tension by means of the eccentric idler shaft See *e* below.

(3) Remove the nuts and wedges on the inside and outside end connections. The most convenient place to disconnect the track is just below the sprocket. When replacing, the link to be added should be inserted midway between the sprocket and the front bogie wheel.

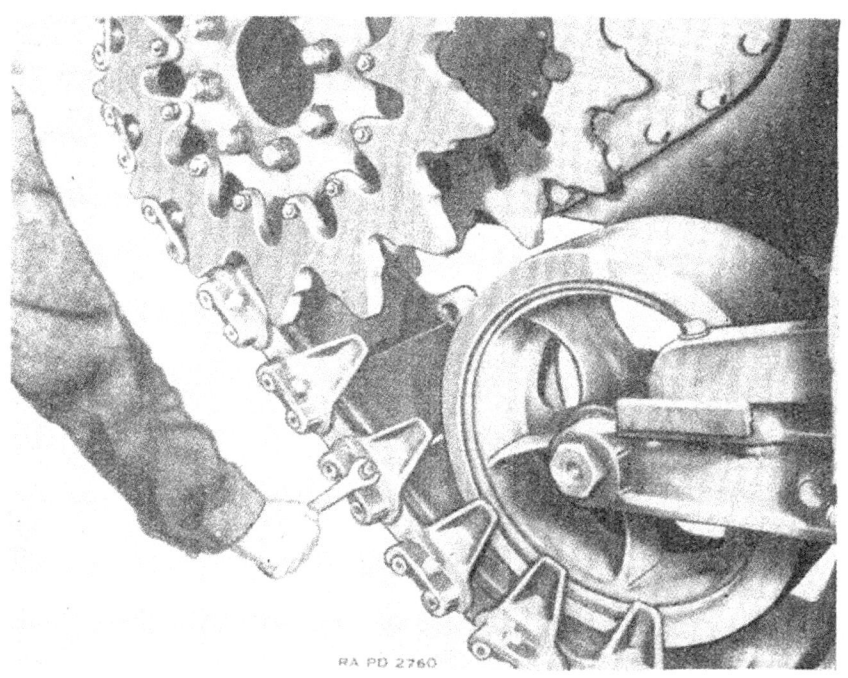

FIGURE 55. Removing end connection from track.

(4) Remove connections as shown in figure 55.

(5) Lower the bottom portion of the track to the ground.

(6) Move the upper portion of the track to the rear over the drive sprocket and supporting rollers by means of a crowbar.

(7) Push or pull the vehicle off the tracks.

d. To replace track (fig. 56).—(1) Lay the track on the ground in a trench at the tank and deep enough so that the bogie can ride onto track, and push or pull the vehicle on it so that the end of the track projects out beyond the front bogie wheel approximateley 16 inches.

(2) Bring the other end of the track over the idler and track-supporting rollers to the drive sprocket.

(3) Take up the slack by revolving the drive sprocket forward with a long bar.

(4) Bring the end shoes together by means of a special jack hook until the guides, wedges, and nuts can be replaced as shown in figure 56.

(5) Adjust track as described below.

e. To adjust track (fig. 57).—(1) Loosen the two outside clamping bolts which hold the spindle of the eccentric cam in the split housing.

(2) Spread the split housing by turning the center bolt counterclockwise.

(3) Raise the clip at the end of the spindle.

(4) Drive the collar (plate) off the serrations of the spindle.

(5) Adjust the idler by turning the hex end of the spindle shaft.

NOTE.—The tension of the tracks should be adjusted as follows in order to prevent throwing of the track: With a straightedge placed on top of the track between the track-supporting rollers, the sag of the track measured from the straightedge should be approximately ¾ to 1 inch, measured between the *front* two support rollers.

(6) When the track is adjusted, slide the collar back onto the serrations of the spindle and lock it with the clip. Release the center bolt by turning it clockwise and tighten the two outside bolts.

98. Idlers.—*a. Description* (fig. 58).—Two large steel idler wheels are mounted on the end opposite the driving sprockets to support the tracks. They are provided with an eccentric adjustment for the purpose of adjusting the tension of the tracks.

b. Adjustment.—See paragraph 97*e*.

c. Lubrication.—A lubrication fitting adaptable to the grease gun is installed in the hub of the idler. It is also equipped with a relief fitting. See section IV, chapter 1, for lubrication instructions.

Figure 56.—Assembly of track.

TM 9-750
ORDNANCE MAINTENANCE

Figure 57 — Idler track adjustment.

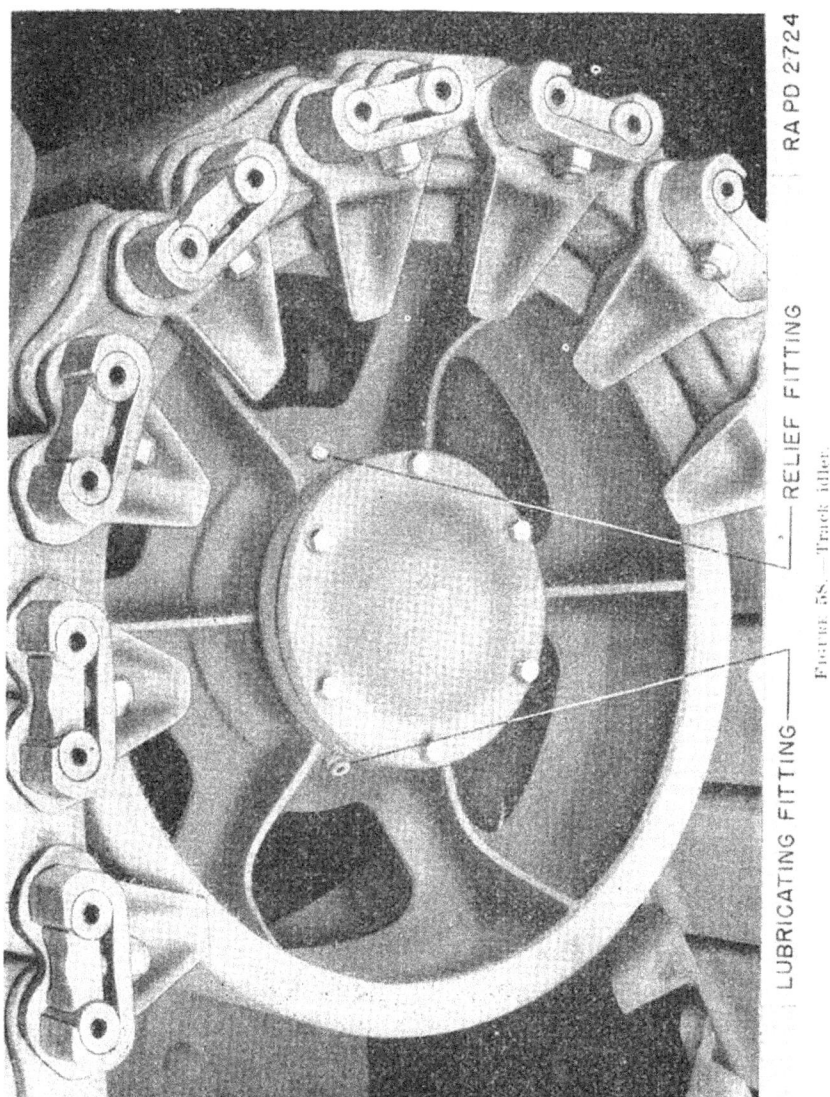

Figure 58. — Track idler.

d. Removing and replacing idler wheels (with track removed).—
(1) Remove the idler cap by removing the six cap screws.

(2) Take out the split pin securing the wheel nut and remove the nut.

(3) Remove the wheel.

NOTE.—Before replacing the wheel, clean bearings, oil retainers, and spacer; when worn or damaged, replace with new parts.

(4) Pack inner and outer bearings with grease.

99. Grousers.—Grousers are provided for conditions of travel where the tracks do not provide sufficient traction such as steep, slippery hillsides, deep mud, or in snow or ice. The grousers are attached to every eighth track shoe by steel pins and a cap screw in the hole provided in the end of the pin in the track block.

SECTION XI

ELECTRICAL EQUIPMENT AND INSTRUMENTS

	Paragraph
General description	100
Battery	101
Battery switch	102
Voltage control	103
Generator filter	104
Magnetic starter switch	105
Instrument panel	106
Maintenance	107
Mounting	108
Fuel gages	109
Solenoids	110
Bow gun firing mechanism	111
75-mm gun firing mechanism	112
Lights	113
Siren	114
Fuse box	115
Radio shielding on conduits and cables	116
Radio	117
Interphone system	118

100. General description.—Due to the increased number of electrically operated accessories, a 24-volt electrical system has been installed in the medium tank. An additional charging plant, described in section XII, equipped with a small motor and generator, has also been installed to supply the extra power and aid in charging the batteries. The wiring diagrams show the electrical circuit of the tank.

101. Battery (fig. 59).—*a. Description.*—Two 12-volt storage batteries are connected in series to maintain the voltage of the system

MEDIUM TANKS M3, M3A1, AND M3A2

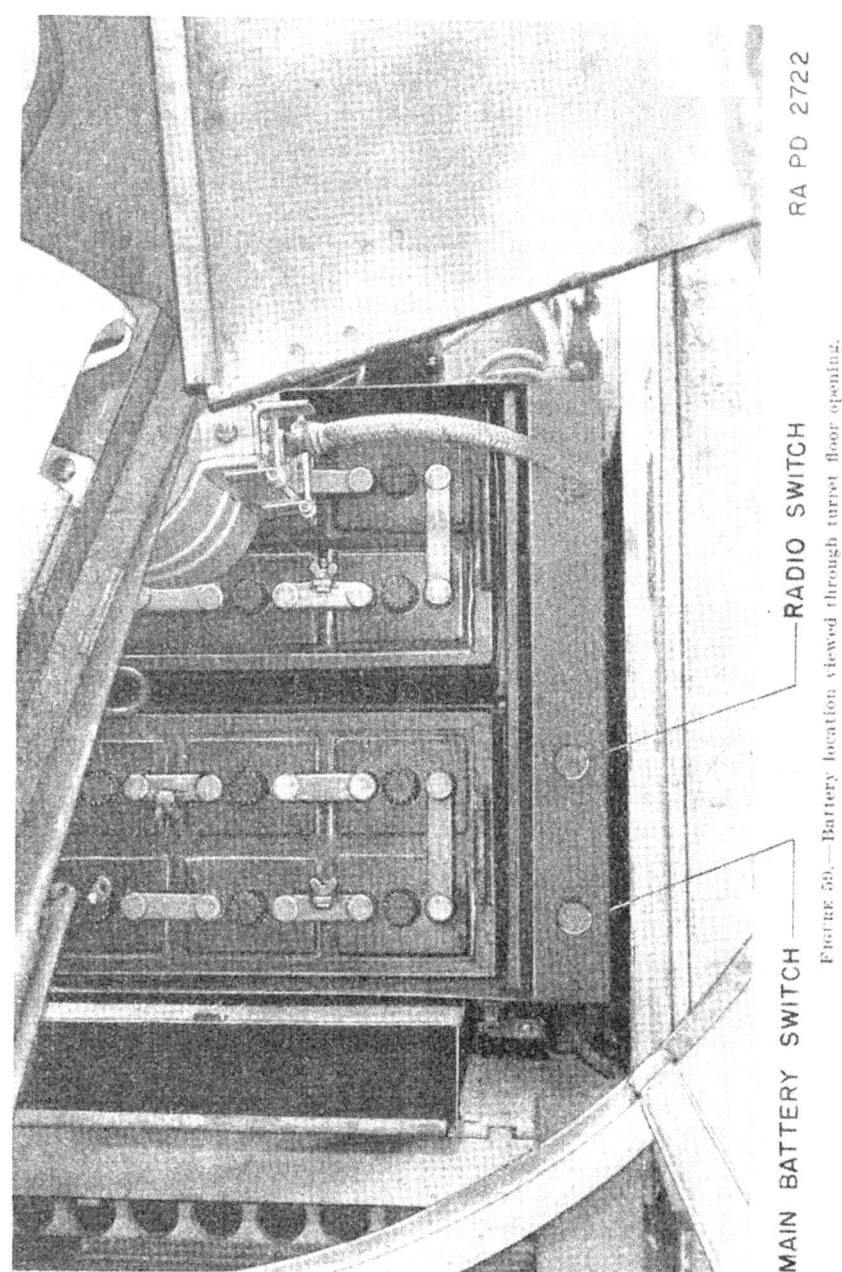

Figure 59.—Battery location viewed through turret floor opening.

at 24 volts. The batteries are installed in the battery compartment, located on the left side of the floor in the fighting compartment, behind the driver, and under the turret basket floor. A direct and separate connection is made between the battery and the radio switch.

b. Checking.—The level and specific gravity reading of the battery fluid should be checked every week and at the end of every long run. Distilled water should be added when necessary to bring the fluid level up to one-quarter inch above the tops of the plates. If the specific gravity of any cell is below 1.200, the battery should be recharged.

c. Removal and replacement.—(1) Open the main battery switch (fig. 59).

(2) Disconnect the battery cables (check legibility of markings).

(3) Remove the bracket hold-down bolts.

(4) Remove the battery through the opening in the turret floor.

(5) Replace in reverse order.

NOTE.—When mounting batteries, see that the positive terminals are next to the propeller shaft, that the posts and the terminals have been cleaned and coated with petrolatum, and that the terminals are tight.

102. Battery switch (fig. 59).—*a.* The battery switch is mounted on the left side of the battery box and is provided to cut off the battery current at its source. A second switch is also provided in the same box for cutting of the radio circuit. To open either circuit, raise the knobs.

b. To replace battery switch.—(1) Remove the partition separating battery and switch compartments.

(2) Remove the switch knob by driving out the retaining pin from the handle.

(3) Remove the terminal nuts and the leads.

(4) Remove the two retaining bolts.

(5) Remove the switch.

(6) Replace in reverse order.

103. Voltage control (fig. 61).—*a. Description.*—To prevent overcharging of the batteries, the generator on the engine and the auxiliary generator are equipped with individual generator control boxes. These boxes of the detachable type, are located near the floor of the fighting compartment, behind the battery box, and include a voltage regulator, current limitator, and reverse current relay.

(1) The voltage-regulating unit maintains the output of the generator at a constant predetermined voltage of 28.4 volts. The current output of the generator is automatically varied in accordance with the state of charge of the battery and the amount of current

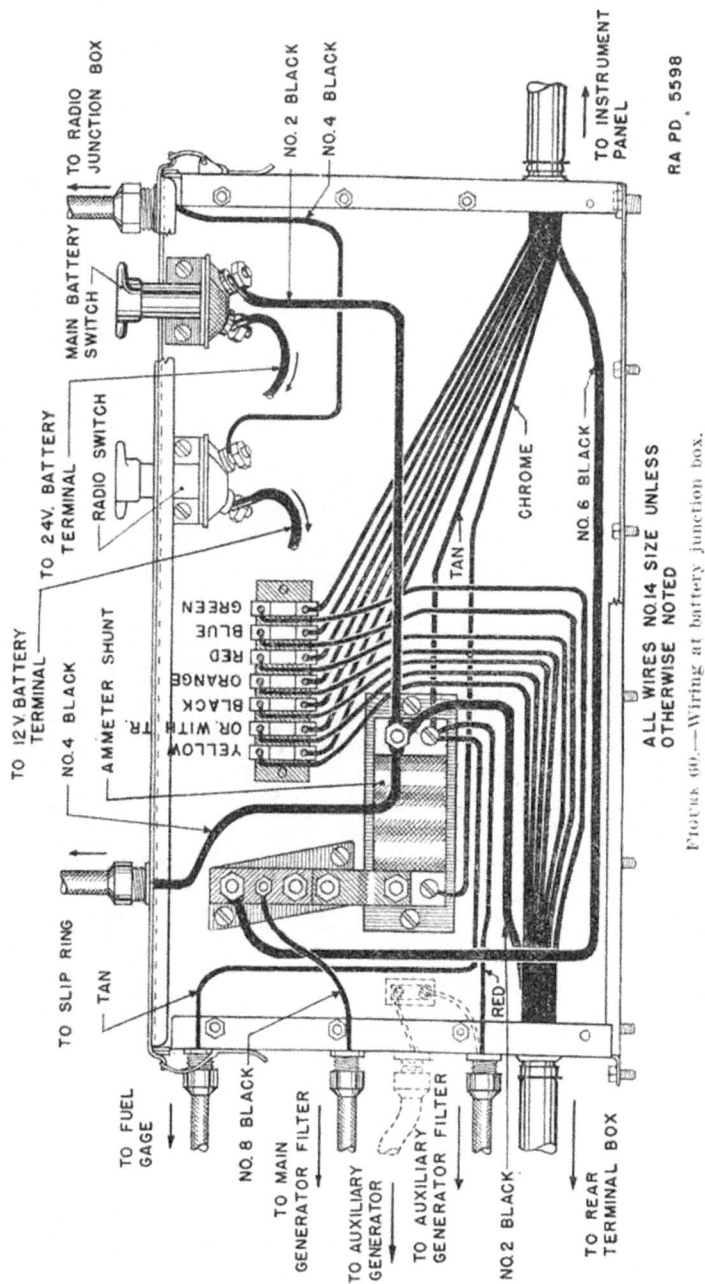

Figure 60.—Wiring at battery junction box.

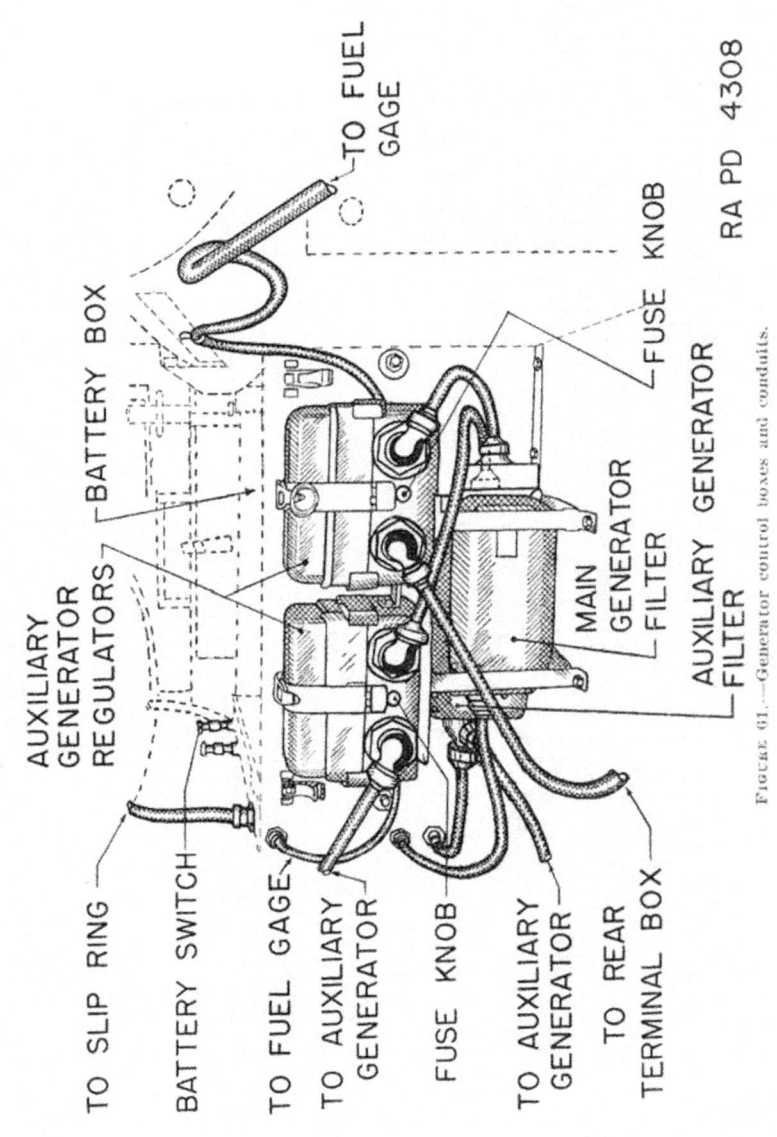

Figure 61.—Generator control boxes and conduits.

being used throughout the vehicle. Thus the proper charge is delivered to the battery at all times without danger of overcharging.

(2) The current limitator unit limits the maximum current output to a value slightly in excess of the rated capacity of the generator.

(3) The reverse current relay or cut-out prevents the battery from discharging through the generator when the generator is at rest or when it is not developing its normal voltage. New models are also equipped with an 80-ampere fuse to further protect against discharging.

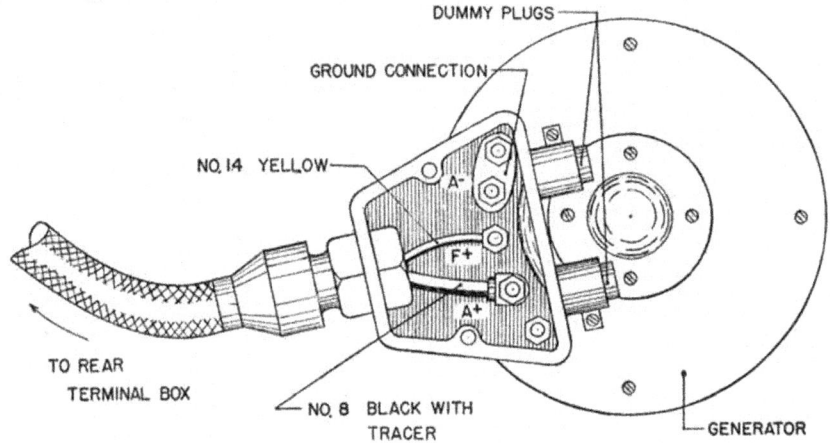

FIGURE 62.—Generator connections.   RA PD 4309

b. Inspection and adjustments.—When properly installed and operated, the generator control units should not require any adjusting. If inspection reveals loose or faulty contacts, improper operation, or if the voltage as indicated by the voltmeter is consistently above or below normal, the unit should be replaced. To replace a unit, release the spring catch, take the top box off the base box, and install a new unit. In pulling the top box from the bottom box, the operation is similar to removing a tube from a radio set.

Caution.—Before attempting to inspect or replace the control box, make sure the main battery switch is open.

c. Lubrication.—No lubrication is required at any time.

104. Generator filter (fig. 61).—A shielded coil is provided in the generator-battery circuits to reduce radio interference. Separate filters are supplied for the generator on the engine and the auxiliary generator. They are located on the floor of the fighting compartment, behind the battery box.

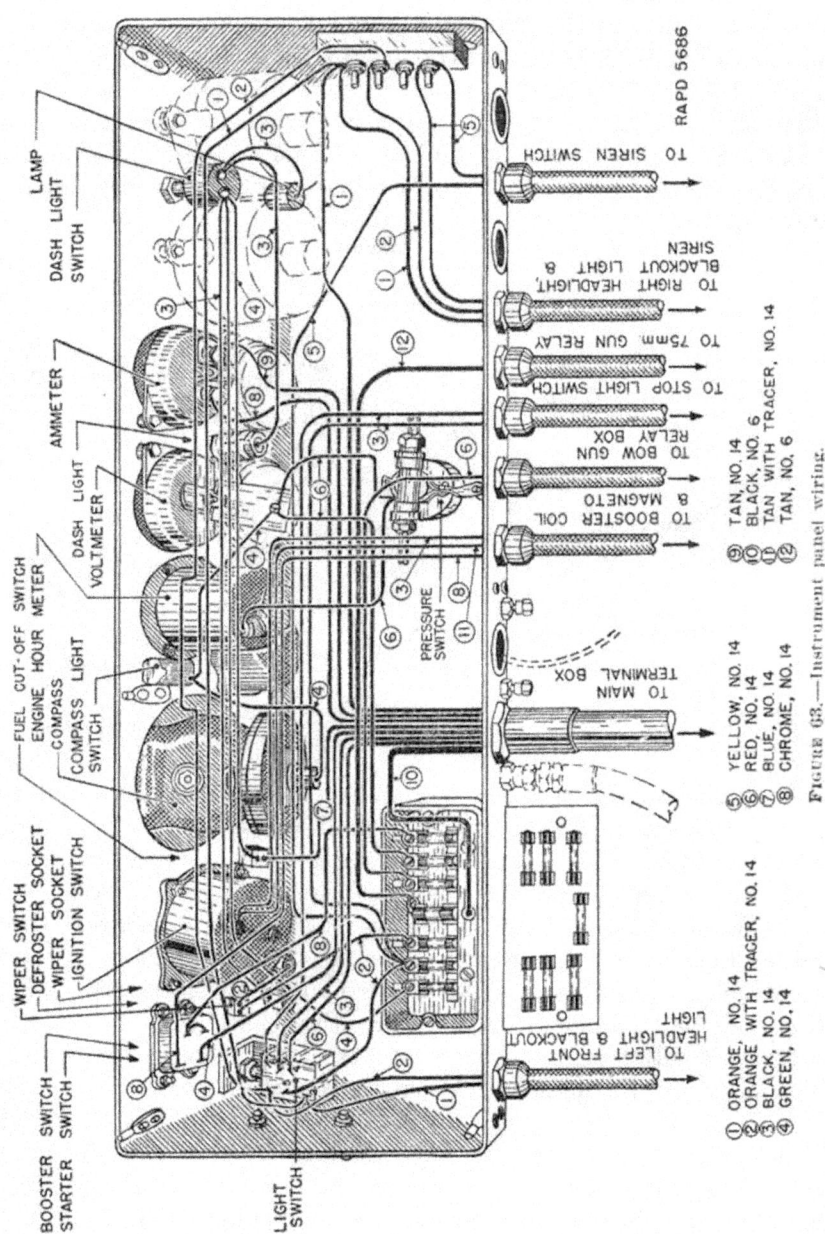

Figure 63.—Instrument panel wiring.

MEDIUM TANKS M3, M3A1, AND M3A2

TM 9-750
105-106

105. **Magnetic starter switch.**—In order to avoid the necessity of running heavy cables, necessary in the battery-starter switch circuit, up to the instrument panel and because of the large current, a magnetic switch is mounted on the engine close to the starter. This switch is actuated by a toggle type starter switch located on the instrument panel (fig. 8).

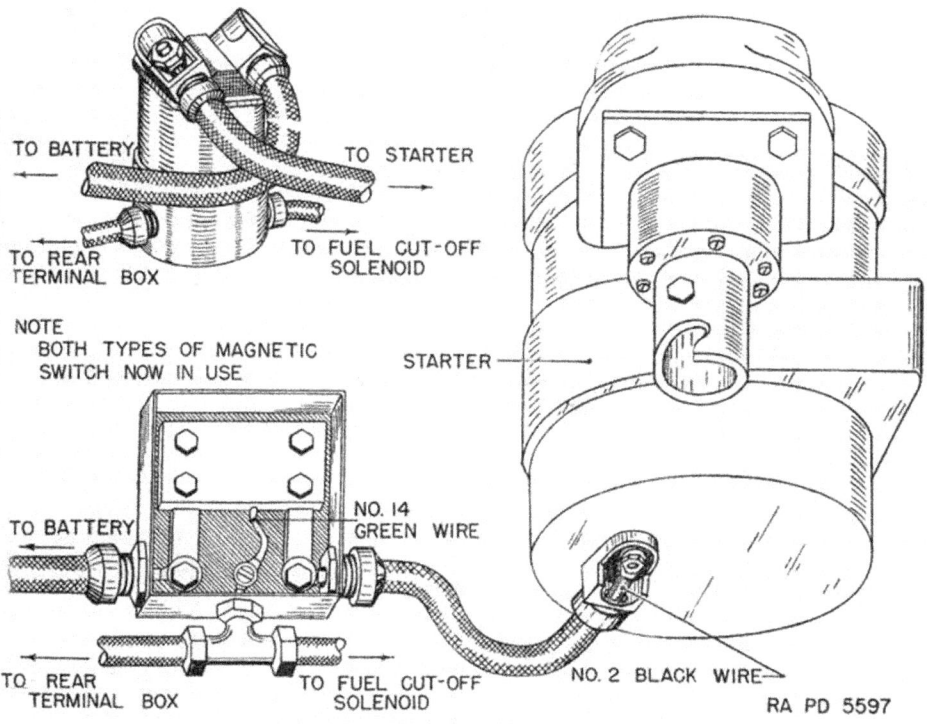

FIGURE 64.—Magnetic switch and starter.

106. **Instrument panel.**—Figures 8 and 63 show the location of the various electrical and nonelectrical instruments, meters, and switches on the instrument panel.

a. Voltmeter.—A voltmeter, located on the instrument panel, indicates the voltage in the circuit. It is connected through a 10-ampere fuse.

b. Ammeter.—The ammeter is located on the instrument panel, and indicates the amount of current in amperes charging or discharging from the battery. The amount of current will vary depending on the engine speed and electrical units in use.

c. Hour-meter.—(1) The hour-meter indicates the total number of hours the engine has been in operation. The meter is installed with a pressure switch. When the required pressure is reached the meter will start operating and will continue to operate whether the engine idles or is under full load until the engine is stopped.

(2) The pressure switch is installed in the oil pressure system by means of a T connection near the oil pressure gage, behind the instrument panel.

d. Ignition switch.—The ignition or magneto switch located on the instrument panel has four positions as indicated on the switch. To extreme right, both magnetos off; to the right of center, right magneto on; to the left of center, left magneto on; to the extreme left, both magnetos on.

e. Blackout switch.—The blackout switch is of the push-pull type. It is located on the upper left-hand corner of the instrument panel. To turn on the surface lights, the spring button on the side of the switch must be pulled out. The switch has four positions. The first is the off position, the second operates the blackout lights, the third operates the service lights, the fourth operates the service stop lights.

f. Starter, booster, compass, fuel cut-off, and wiper switches.—These are all toggle switches located on the instrument panel, and so labeled.

g. Dash light switch.—The dash light switch is of the push-pull type, located on the lower right section of the instrument panel.

h. Defroster and windshield wiper sockets.—The defroster and windshield wiper sockets are located in the upper left corner of the instrument panel.

i. Tachometer.—The speed of revolution of the engine crankshaft is indicated by the tachometer. It is driven by a flexible, encased shaft connecting to the tachometer drive in the rear face of the accessory case. Tachometers that register incorrectly will be replaced.

j. Speedometer.—The speedometer is driven by a flexible shaft connecting from the spiral gear on the output shaft of the transmission. An adometer is also incorporated in the head of the speedometer.

k. Oil pressure gage.—This gage records the pressure in the engine oil manifold through a ¼-inch copper tube connected to the left side of the oil pump housing.

l. Oil temperature gage.—This gage indicates the engine oil temperature by means of a temperature-sensitive device located on a fitting in the oil pump.

m. Fuel primer.—The fuel primer is located on the instrument panel and is discussed in section III of chapter 1 and section IV of chapter 2.

n. Compass.—(1) *Description.*—A compass is installed in some of the tanks. Necessary illumination, controlled by a tumbler switch to the right of the compass, is provided by a small lamp. The bulb is accessible for replacement by removing the upper knurled plug. The two-prong socket at the rear end permits the removal of the compass without disturbing the conduit fastening.

(2) *Compensation.*—(*a*) In order to correct or compensate for the attraction of the various metal components near the compass, a means of compensation is provided within the instrument. By removing the two screws, the upper lamp holder or shield may be removed, thus exposing the compensating screws. Using the small brass screw driver provided for this purpose, turn the screws until the white dot in the screw slot alines with the white dot on the compass body.

(*b*) With all equiment of a magnetic nature in place, head the vehicle due magnetic north, as determined by an instrument outside and away from the vehicle. (A surveyor's transit may be used.) Turn the "N–S" screw until the compass reads "N". Head the vehicle due west and set the compass at "W" by turning the screw marked "E–W". Head the vehicle due south and remove *one-half* the existing error by turning the "N–S" adjusting screw. Head the vehicle due east and remove *one-half* the existing error by turning the "E–W" adjusting screw.

(*c*) Recheck by heading the vehicle on the magnetic headings shown on the compensating card and record the corresponding compass readings in the spaces provided.

(*d*) The compensating card is carried in the deviation card holder, attached to the panel above the compass.

(3) *Maintenance.*—At frequent intervals, inspect the instrument for the appearance of bubbles in the bowl, and if necessary remove the filler plug and refill with ethyl alcohol. Compensation for error due to variable magnetic influences should be made whenever such conditions arise.

107. Maintenance.—All instruments and switches that become inoperative should be exchanged for serviceable ones.

108. Mounting.—To replace any instrument or switch from the instrument panel—

a. Open the main battery switch.

TM 9-750
ORDNANCE MAINTENANCE

b. Remove bottom cover of box by removing six nuts.

c. Remove screws on face of panel holding the defective instrument or switch and the nuts, spacer, and lock washers.

d. Remove the defective instrument or switch from the bottom of the instrument panel.

e. Replace in reverse order.

109. Fuel gages (fig. 38).—The fuel gages are electrically operated. They are located just above the power tunnel close to the rear bulkhead of the main fighting compartment.

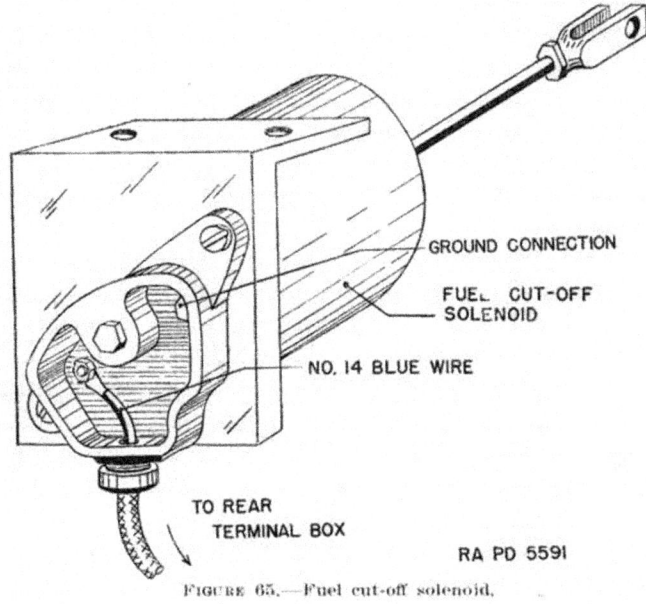

FIGURE 65.—Fuel cut-off solenoid.

110. Solenoids.—*a.* The firing of the combination 37-mm gun and the caliber .30 machine gun, the bow guns, and the 75-mm gun is accomplished by solenoids which operate the firing levers. The switches controlling the electrical circuits to these guns were conveniently located for the respective operators. The solenoids function automatically upon operation of the control switch.

b. When properly installed, these solenoids should not require any attention. If trouble is experienced during normal operation, a heavy jumper placed across the contact terminals will indicate whether the solenoid is inoperative or trouble lies in some other part of the circuit. If the solenoid is inoperative, replace with one that works.

111. Bow guns' firing mechanism.—*a.* The switches which control the firing of the bow guns are located on the steering levers. Either switch will energize the solenoid and fire both guns.

b. The electric power is supplied by the main batteries and can be cut off by a push-pull switch located in the relay box (fig. 66). The relay box is attached to the front hull plate to the left of the instrument panel, directly above and between the bow guns.

c. The fuse for the bow gun circuit is in the main fuse box in the instrument panel.

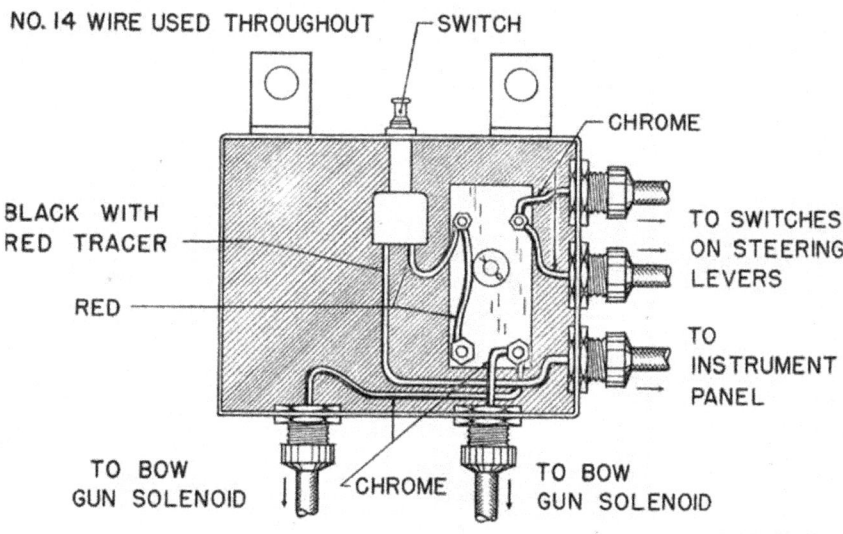

FIGURE 66.—Bow gun relay box.

112. 75-mm gun firing mechanism (par. 35*a* (7)).—The wiring diagram for the electric firing mechanism is shown in figure 67. To fire the gun, depress the firing button located at the center of the traversing handwheel (fig. 14). Depressing the button actuates a switch which sends an electric current through a solenoid. The solenoid plunger pushes against a shaft that operates the firing plunger lever. The switch box (fig. 67), which contains a relay and circuit breaker, is mounted on the left side of the gun. The fuse for the 75-mm gun circuit is in the main fuse box.

113. Lights.—*a. Description.*—(1) Two headlamps, two front blackout lamps, and two rear lamps are provided. The tail lamps are combination lamps. The rear left is a combination stop, service, and blackout lamp. The lower lens is the service and service stop light and takes a double beam 3–21 candlepower bulb. The upper

TM 9-750
113 ORDNANCE MAINTENANCE

lens is the blackout light. In the rear right lamp, the lower lens is the blackout light and the upper lens is the blackout stop light.

(2) Both stop lights are controlled by stop light switches connected with the steering hand levers. No stop signaling will be seen until both hand levers are pulled, indicating a slowing down or full stop.

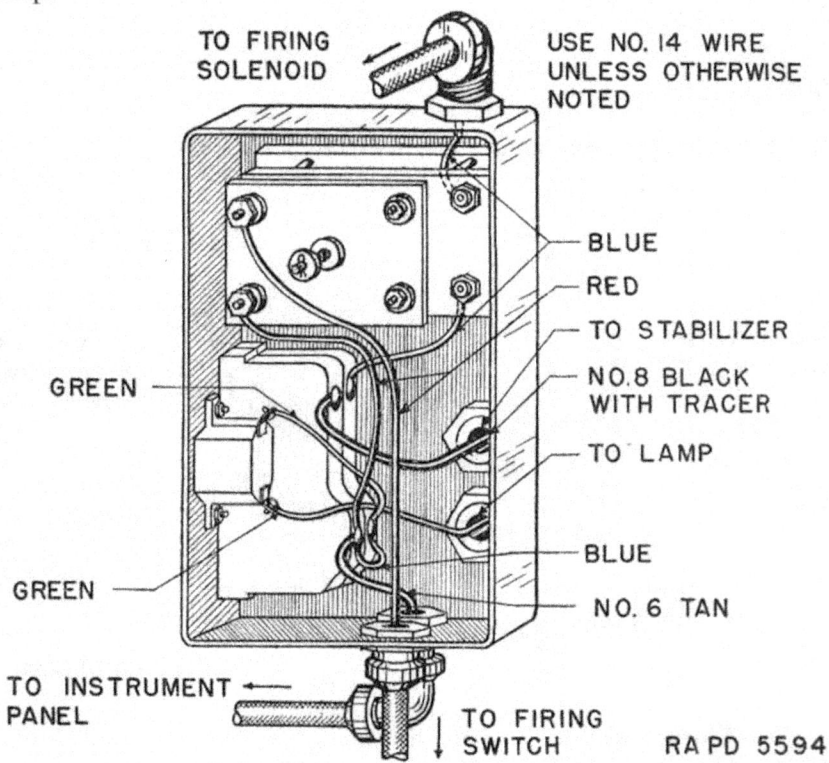

FIGURE 67.—75-mm gun switch and relay box.

b. Focusing.—The headlights are focused by means of a focusing screw back of the headlamp. The trunnion and pivot mounting brackets provide for altering their position.

c. Bulbs or lamps used.

Type	Candlepower	Contact
Headlights:		
2 service	40	Single.
2 blackout	3	Single.
Rear right:		
1 blackout stop light	3	Single.
1 blackout tail light	3	Single.

Type	Candlepower	Contact

Rear left:
1 service tail and stop light
  (upper unit) _____ 32   Double (double filament).
1 blackout tail light _____ 3   Single.
2 instrument panel lights _____ 3   Single.
2 compass lights (1 spare) _____ 3   Single.
1 75-mm gun light _____ 3   Single.
2 turret lights _____ 3   Single.
1 fuel gage light _____ 3   Single.

114. Siren.—The siren is located on the front right fender of the tank and is operated by a foot button at the driver's left foot. To replace siren—

a. Open the main battery switch.

b. Disconnect the electric cable at fender connector.

c. Remove the nuts and bolts.

d. Remove siren.

e. Replace in reverse order.

115. Fuse box.—*a. Description* (fig. 63).—A fuse block mounting seven fuses is inclosed in the fuse box, located in the left end of the instrument panel box. The fuse box is stamped with the amperage of the fuse required in each position. A duplicate set of fuses is clipped to the cover of the fuse box for replacement purposes. These should be clipped in places as shown. Spares should be replaced as soon as possible after being used.

NOTE.—Only the fuse designated will be installed. Larger, smaller, or makeshift fuses will not be used. The fuse box cover can be removed from the under side of the panel by removing the two machine screws. Fuses not in the main fuse box are shown in their respective circuits.

b. Fuse sizes and location.—The correct fuse installation in the fuse block is as follows, listed from the left side of the tank:

Amperage	Instrument
20	Dash light and fuel cut-off.
20	Siren and lights.
20	Wiper switch.
60	75-mm gun.
30	Bow guns.
10	Voltmeter, compass, and hour-meter.
10	Booster, starter, and ignition.

c. Fuse block assembly replacement.—(1) Open main battery switch.

(2) Remove two screws on back of fuse block assembly and remove cover.

(3) Remove four screws from the back of the instrument box.

(4) Disconnect all wires at connections in fuse block and pull wires through.

(5) Remove fuse block assembly.

(6) Replace in reverse order.

116. Radio shielding on conduits and cables.—*a. Description.*—Both high- and low-tension wiring is shielded by flexible shielded conduit. The radio shielding required for the engine and its mounted accessories is formed in a unit known as a "harness" and is described in section III. The flexible shielded conduit is oil- and water-resistant. Within the shielded conduit, standard unshielded ignition or primary automotive cable is used.

b. Maintenance and inspection.—(1) The electrical wiring, the shielding, and conduits require frequent inspection and check.

(2) All crushed shielding and conduits should be replaced. Oily or dirty spark plug shields or shielding fittings should be cleaned and all coupling nuts tightened. Wires and conduits which have become oil-soaked should be replaced.

(3) In cleaning couplings or spark plug shields, a solution of carbon tetrachloride should be used. If carbon tetrachloride is not available, dry-cleaning solvent may be used. After cleaning and drying, the threads of each coupling should be cleaned with a small wire brush to remove high-resistance oxidation which sometimes forms on the inside of aluminum couplings.

117. Radio.—*a. Description.*—There are two types of radio installed. The majority of tanks will use the tank-to-tank model, operating on 24 volts, which will be installed in the left sponson. The staff and command tanks will have a radio set in the command channel in addition to the other set. The command model will be installed in the rear right sponson.

b. Radio junction box.—A radio take-off box is located just above and in front of the main battery switch, which is attached to the left side of the battery box.

c. Radio switch.—A switch is provided to cut off the battery current for the radio at its source. It is located in the main battery switch box, in the left side of the battery box. To cut off the current, the knob is raised.

118. Interphone system.—The model RC53A interphone communication set is installed to provide a means of enabling the tank crew to communicate with one another. There are earphone facilities for all members of the crew so that they may hear orders from the tank commander, and equipment for five of the crew—all but the 75-mm and 37-mm gun loaders—to speak to the other crew members by use of telephones.

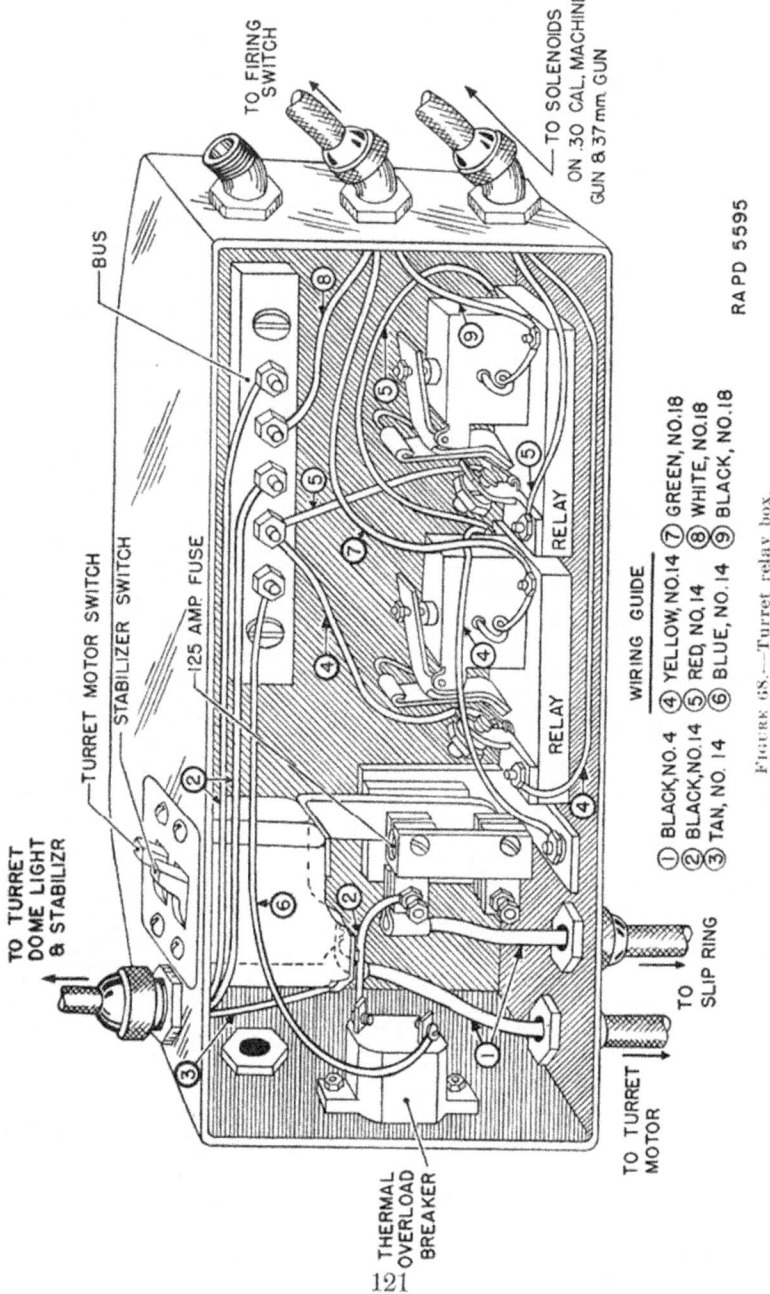

Figure 68.—Turret relay box.

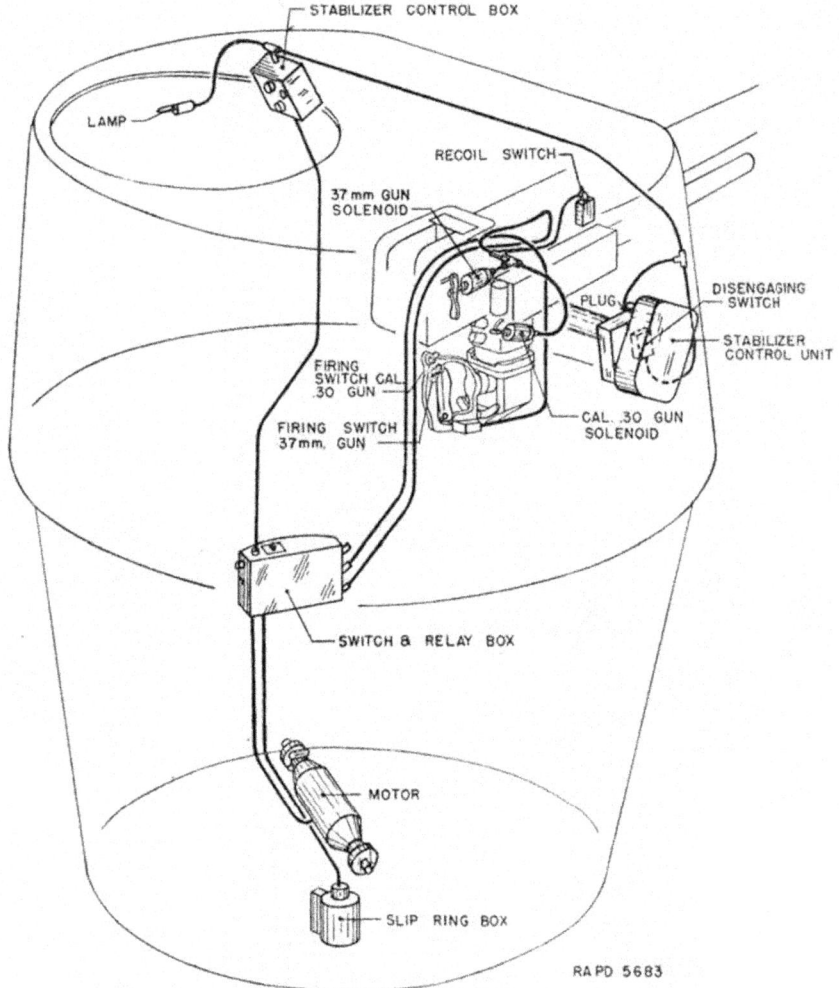

Figure 69.—Turret conduits.

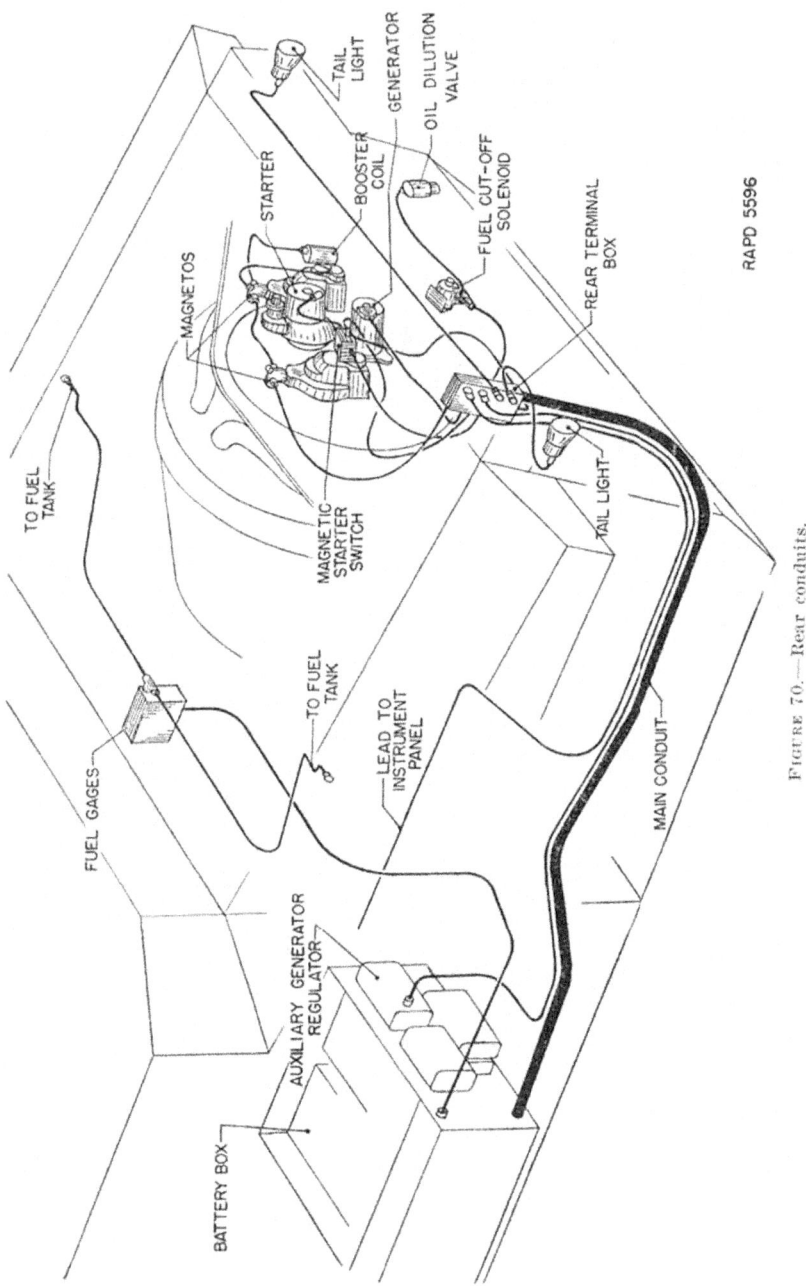

FIGURE 70.—Rear conduits.

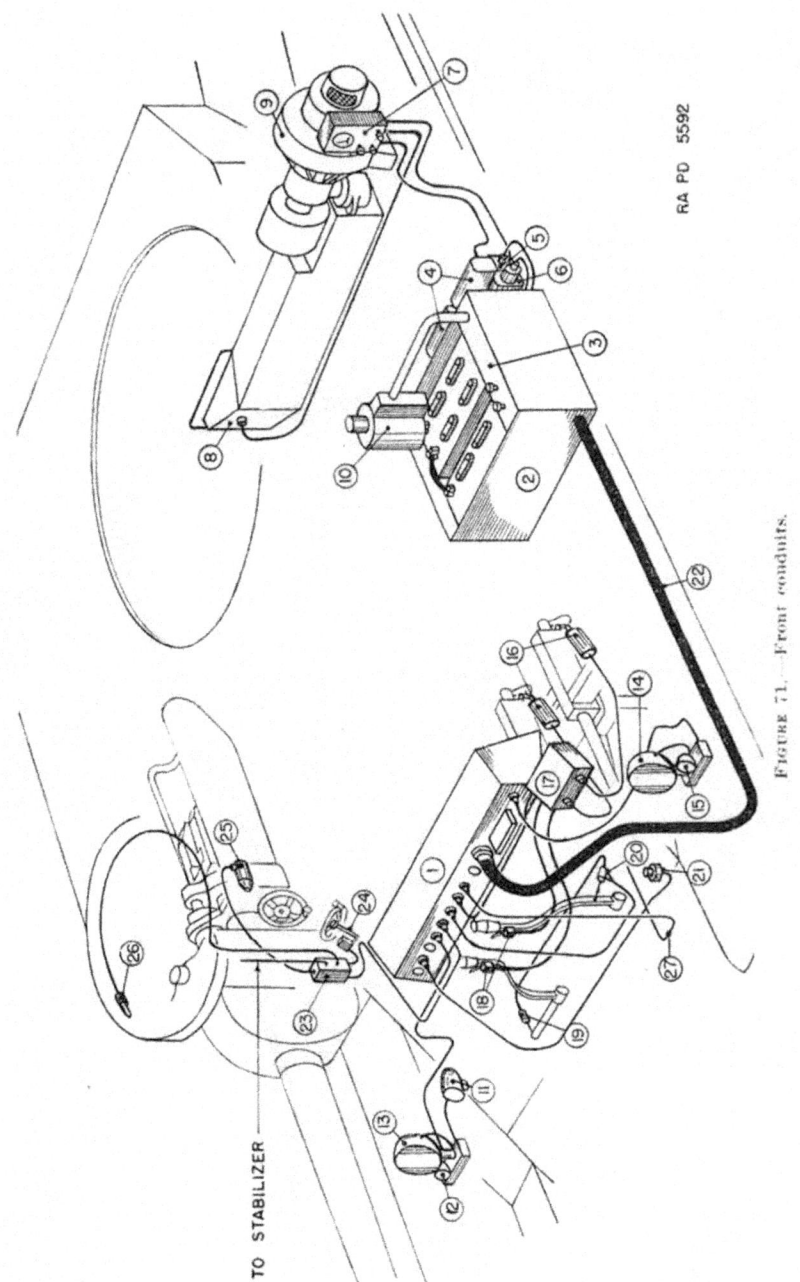

Figure 71.—Front conduits.

MEDIUM TANKS M3, M3A1, AND M3A2   TM 9-750
                                   118

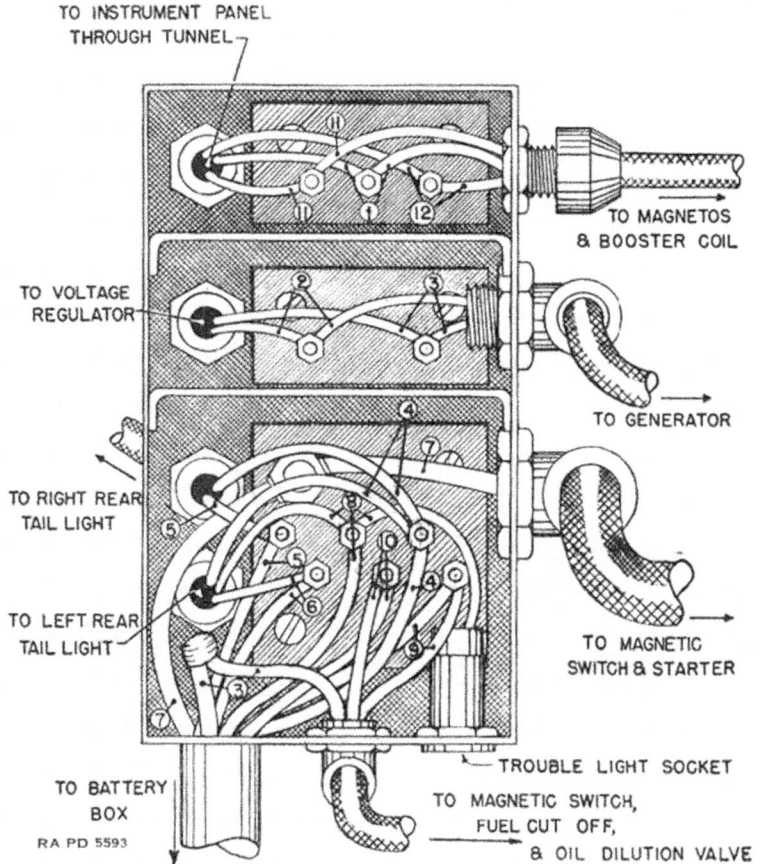

WIRING GUIDE

1. Chrome, No. 14.
2. Black with tracer, No. 8.
3. Yellow, No. 14.
4. Orange, No. 14.
5. Red, No. 14.
6. Black, No. 14.
7. Black, No. 2.
8. Orange with black tracer, No. 14.
9. Blue, No. 14.
10. Green, No. 14.
11. Tan, No. 14.
12. Black with tracer, No. 14.

FIGURE 72.—Wiring at rear terminal boxes.

125

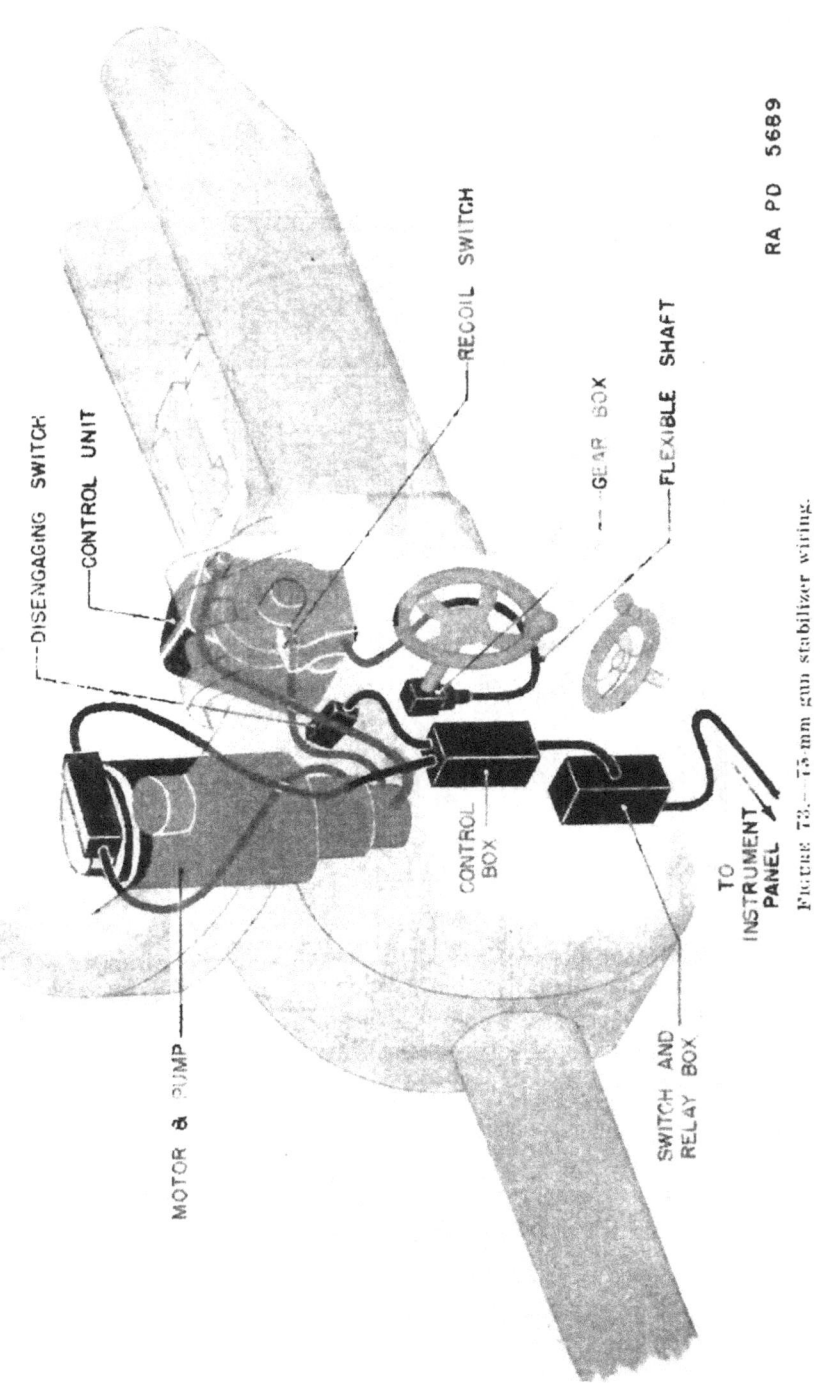

Figure 73.—75-mm gun stabilizer wiring.

MEDIUM TANKS M3, M3A1, AND M3A2

TM 9-750
119-120

Section XII

AUXILIARY GENERATING UNIT

	Paragraph
General	119
Description	120
Maintenance	121

119. General (figs. 74 and 75).—*a.* The auxiliary generator is a self-contained generating unit for charging the tank batteries and for preheating the tank engine compartment in cold weather by means of heat from an electric heating element and heat from the generator engine. It may also be used for heating the crew compartment. The unit is located in the left rear corner of the fighting compartment on a level with the turret deck.

b. It is desirable that the auxiliary generating unit be in operation whenever the turret is being power-operated (a 24-volt electric motor drives the hydraulic pump) or when the guns are being fired (all guns are fired through solenoid switches, excepting the caliber .30 machine gun in the cupola).

120. Description.—*a.* The unit consists of a 2-cycle, 2-port, single-cylinder, air-cooled, inverted type gasoline engine, operating at 3,500–3,600 rpm; a 30-volt, 1,500-watt generator; a blower; and a heater duct which runs from the unit to the engine compartment.

b. The control box contains the three control buttons, marked "start," "battery," and "heater." When the "battery" button is pressed, the generator is used to charge the battery. When the "heater" button is pressed, the generator output is used in the heater element to heat the engine or crew compartment. The "start" button is used for starting the unit electrically. The ammeter indicates either the charging rate of the generator or the current in the heater circuit.

Caution.—Heater button should be pushed in at all times unit is operating except when battery needs charging.

c. The fuel tank, located in the left sponson, holds approximately 2½ gallons of fuel. The fuel consists of a mixture of oil and gasoline. Use ⅜ pint of engine oil, SAE 50 or 60, mixed thoroughly with each gallon of gasoline. Lubrication for the entire engine is obtained by mixing the oil with the gasoline and it is important that the oil be thoroughly mixed with the gasoline before pouring into the tank.

d. To operate the unit.—(1) For cold weather starting, below 32° F.—

(*a*) Close the carburetor choke. (The choke is open when the lever is against the stop pin.)

(*b*) Depress the starting button on the control box. Release the button as soon as the engine starts and immediately open the choke partially, easing to full open position as the engine warms up.

(2) For warm weather starting, above 32° F.—

(*a*) Depress the starting button on the control box. Release the button as soon as the engine starts.

(*b*) Do not use the choke unless the engine does not start within 5 seconds. If the engine does not start, use the choke as in previous instructions.

NOTE.—Do not use the choke as a throttle. The automatic governor keeps the engine operating at the proper speed at all loads.

(3) To stop the engine, press the red stop button and hold firmly until the engine stops. Close the fuel line shut-off cock beneath the fuel tank. The cock is closed when it is turned to the full clockwise position.

(4) To heat the crew compartment, after starting the unit and operating on the heater circuit, open the heater duct door and fold it back along the top of the duct so the handle fastens in the catch. To reclose the door, unfasten the handle and slide it down into position.

(5) A manual method of starting in an emergency is incorporated in the unit. To do this pry open with a screw driver the three clips holding the shield over the magneto and remove the front half. Wind the starting rope counterclockwise on the starting plate, which is then exposed. Use the choke as previously described. Pull the rope hard, giving the engine a quick spin. Repeat if necessary until the engine starts.

121. Maintenance.—*a.* The maintenance requirements are as follows:

(1) *25 hours.*—(*a*) Replace spark plug. Check gap for 0.025-inch clearance.

(*b*) Remove and clean both sides of spark plug baffle.

(*c*) Take air filter apart and rinse with dry-cleaning solvent. Dip upper end of screen in SAE 30 engine oil. Reassemble.

(*d*) Clean strainer in fuel inlet connection on top of carburetor bowl.

(*e*) Clean strainer in top of fuel filter sediment bowl.

(2) *100 hours.*—(*a*) Replace spark plug baffle. Use new gasket.

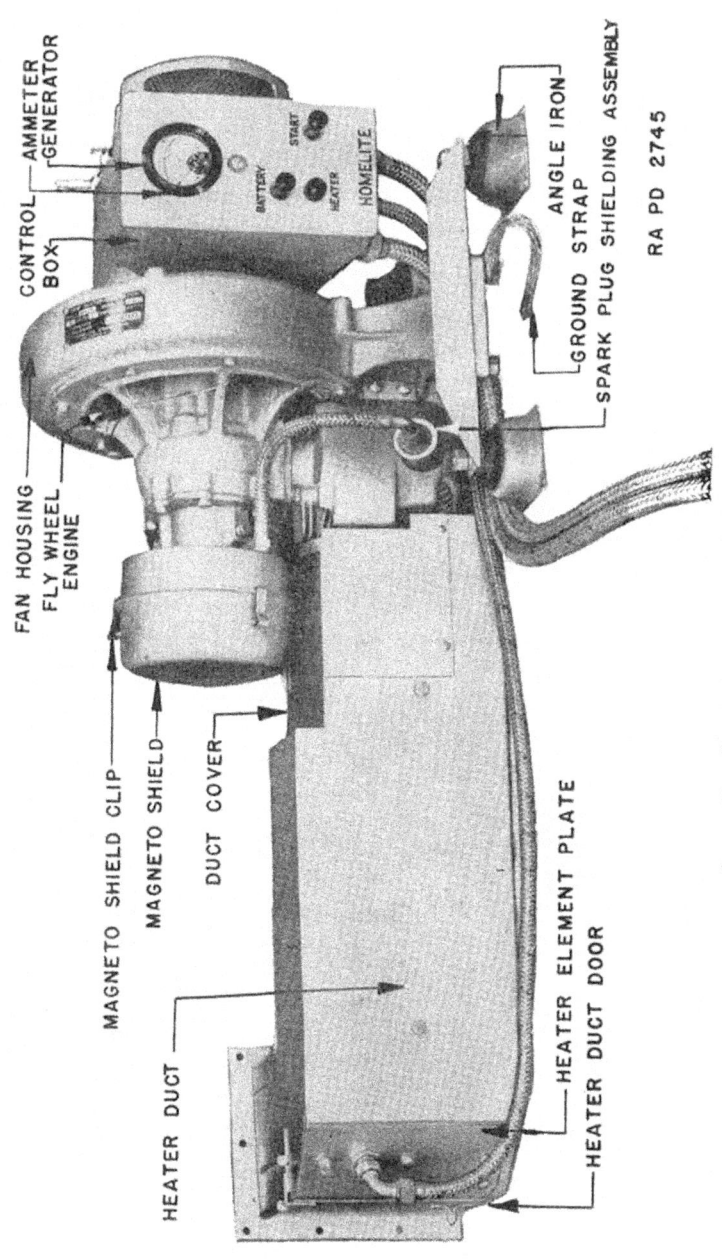

Figure 74.—Front view of auxiliary generator unit.

TM 9-750
121

ORDNANCE MAINTENANCE

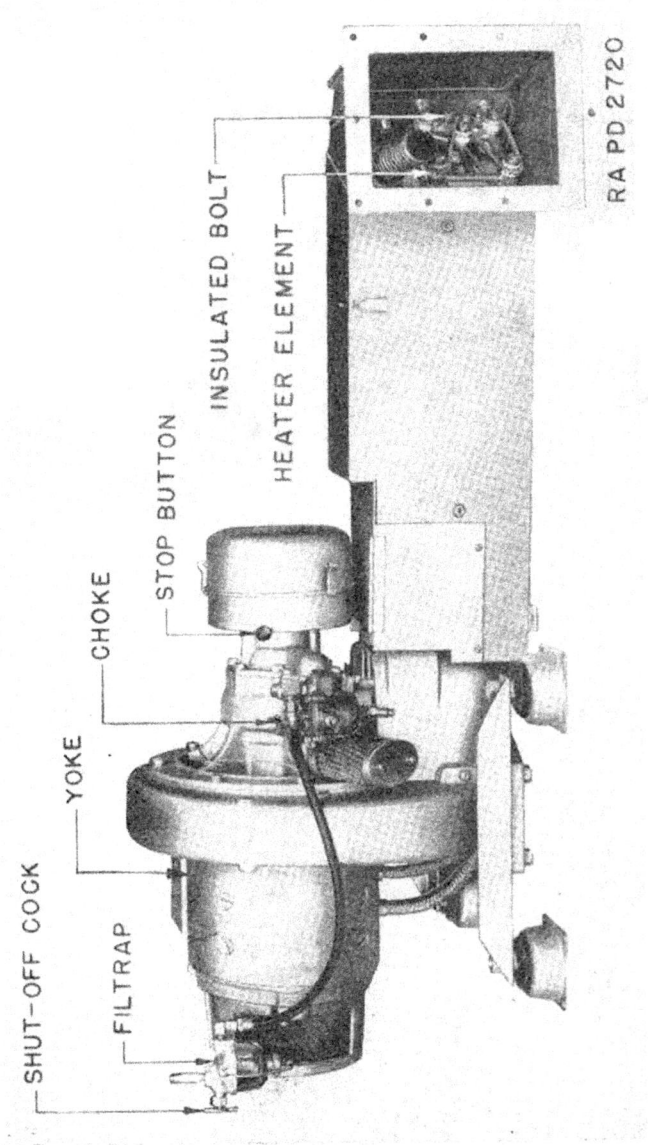

Figure 75.—Rear view of auxiliary generator unit.

MEDIUM TANKS M3, M3A1, AND M3A2

TM 9-750
121

(*b*) Check cylinder for carbon. If carbon is forming from top of exhaust port, O. K.; if from bottom, clean thoroughly. Remove spark plug to scrape cylinder head.

(*c*) Check magneto points to see that gap is exactly 0.020 inch. If points are uneven or pitted, replacement is recommended. If replacement is not available, use smooth carborundum stone to true them. Remove dust with dry cloth. Never use a file on contact point surfaces.

NOTE.—Carburetor is adjusted correctly (1½ to 1¾ turns on needle valve) when it will run about 30 seconds with full choke. (Carburetor adjustments by ordnance personnel only.)

Set generator brushes (4) for a 50-ampere reading.

Set auxiliary on heater switch (unit will then operate under full load) when making adjustments on points, etc.

When starting with battery don't use choke unless cold or engine fails to start in 5 seconds. When starting by hand use choke.

The engine averages 2½ hours operation to the gallon of fuel.

Battery is fully charged if ammeter reads 3 to 5 amperes. Battery is in discharged condition if ammeter reads 50 amperes.

No reading should show in main ammeter on instrument panel when starting auxiliary generating unit.

b. Spark plug.—(1) To replace the spark plug, remove the cap from the spark plug shield and take out the plug with a spark plug wrench. In removing the plug, the baffle may come out. If it does, remove it from the plug. Always use a Champion J-10 Commercial spark plug or equivalent. Note that in replacing the plug the copper gasket goes outside of the baffle surrounding the plug.

(2) Each time the plug is removed, the baffle should be taken out for cleaning. It is extremely important to scrape out and thoroughly clean both sides of the baffle. However, the baffle should be replaced every 75 hours of operation of the unit. Always use a new gasket when replacing the baffle.

c. Magneto.—To check magneto points, proceed as follows:

(1) Pry open the three clips on the shield over the magneto with a screw driver, and take off the front half.

(2) Remove the magneto rotor by loosening the rotor nut. The contact points are then exposed.

NOTE.—Do not remove the three screws holding the starting plate to the rotor.

(3) Remove the spark plug as in the instructions above, to relieve compression and permit turning the flywheel.

(4) Rotate the flywheel until the gap reaches its maximum opening. Check gap with feeler gage.

(5) If it is necessary to adjust the gap to 0.020 inch, *loosen the screw slightly* which fastens the contact point assembly to the starter plate and move the entire breaker mechanism toward the cam to increase the gap, or away from the cam to decrease the gap.

(6) Tighten the contact point assembly, fastening screw securely and recheck the gap with a feeler gage. Readjust if necessary. Tightening of the setscrew sometimes changes the adjustment.

(7) The entire contact point assembly pivots on the breaker lever bearing pin, which permits adjustment of the gap without altering the relationship between the contact point surfaces. If the breaker cam is removed from the intake valve shaft, replace with the arrow (indicating rotation) on the outside.

(8) Uneven or pitted contact points may be restored to a true, even condition by the use of a smooth carborundum stone, after which all dust particles should be removed with a dry cloth. However, if points are in this condition a new set is recommended. *Do not use a file on the contact point surfaces.* Stiff paper or cardboard will remove the oxide formation which results from long idleness.

d. Carburetor.—Adjustment of the carburetor will be made by trained ordnance maintenance personnel only.

(1) Keep the strainer in the fuel inlet connection on top of the carburetor bowl free from sediment. When this strainer is being inspected, open the fuel line shut-off cock beneath the fuel tank to make certain there is a free flow of fuel to the carburetor.

(2) If the fuel does not flow freely remove the fuel filter sediment bowl beneath the fuel tank and clean the strainer in the top of the bowl.

e. Air filter.—Clean the air filter monthly on the carburetor intake. Take apart and rinse with dry-cleaning solvent. Then dip the upper end of the screen in SAE 30 engine oil, and reassemble.

f. Removing unit from tank.—For complete overhaul, or if unit fails to operate after minor repairs, remove it from the tank. To remove, proceed as follows:

(1) Open the main battery switch of the tank.

(2) Remove the duct cover.

(3) Remove exhaust coupling.

(4) Close the shut-off cock on the fuel filter and disconnect the fuel line at the carburetor.

(5) Remove the control box from the generator as follows:

(*a*) Take off the three black knurled buttons and lock washers. Remove the screw below the ammeter and take off the cover. Allow the cover to hang by the inside connecting wires.

MEDIUM TANKS M3, M3A1, AND M3A2

TM 9-750
121-123

(*b*) Disconnect at the terminals in the control box the three wires that come from the generator yoke. These are coded red, red and green, and black. The wires can be removed with a pair of pliers.

(*c*) Remove the two screws holding the control box to the generator yoke.

Caution.—Use care not to drop these screws and washers in back of the panel board. If they are accidentally dropped, be certain to remove them before assembling, as they may cause a short circuit.

(*d*) Push the three wires disconnected in accordance with (*b*) above through hole in box and take off the control box.

(6) Remove the four bolts which fasten the fan housing to the steel angles.

(7) Take off the nut on the front end of the right angle and swing it away from the unit to facilitate removal.

(8) Lift out the unit.

(9) If it is necessary to take out the duct assembly, unscrew the nut on the heater mounting plate, and remove the heater assembly.

(10) Disconnect the flange at the end of the muffler by taking out the three screws accessible through the duct door and heater element plate opening.

(11) Take out the screws holding the duct to the engine bulkhead and sponson and remove the duct.

g. Reinstallation of unit in tank.—The unit may be installed by reversing the instructions given above. Make sure that gaskets are used in the connections between the duct and the engine bulkhead and between the cylinder and flexible exhaust coupling.

SECTION XIII

STABILIZERS

	Paragraph
General	122
Inspection	123
Trouble shooting	124
Charging with oil and removing air from system	125
Replacement of assemblies	126

122. **General.**—Figures 76 and 77 show the complete assemblies of the stabilizer systems. Repairs must not be made to individual assemblies, but rather the assemblies will be replaced as units. Corresponding assemblies for the two systems are similar.

123. **Inspection.**—*a. 25-hour inspection.*—After every 25 hours of operation both stabilizer systems should be checked as follows:

(1) Check for free motion between the control unit and the hand-

wheel and adjust the backlash between the worm gear and worm wheel if necessary.

(2) Check for free movement in the mounting of the piston and cylinder assembly.

(3) Tighten the packing gland on the oil pump shaft by removing the cover plate and turning the packing gland nut in a clockwise direction with a screw driver. Do not get it too tight.

(4) Lubricate the four alemite fittings shown in figure 80.

(5) Check all external electrical connections to see that they are tight.

(6) Check all oil line connections.

(7) Check for presence of air in the oil system.

b. Recoil switch adjustments.—The adjustment of the recoil switch must be checked periodically to insure its proper functioning. The switch is so constructed and mounted that its contacts are held open when the gun is in battery. When the gun is in battery position the switch plunger should not protrude more than one-eighth of an inch. It must be adjusted so that the contacts will close with the first recoil movement of the gun, but the contacts should not close until the gun is fired. The switch is adjusted by bending the clip which contacts the plunger.

124. Trouble shooting.—Troubles most frequently encountered and their probable causes are as follows:

Trouble	*Probable cause*
Hunting	Improper stiffness adjustment.
	Excessive friction in trunnion bearings.
	Air in system.
Loss of sensitivity	Improper stiffness adjustment.
	Improper grade of oil.
	Excessive friction in trunnion bearings.
	Air in system.
	Unbalance of gun.
	Play between control unit and handwheel.

125. Charging with oil and removing air from system.—When charging the system with oil it is very important that all air that may be trapped in the system be removed for proper operation of the stabilizer. Therefore, the following procedure must be adhered to:

a. Throw the master switch to the "off" position.

MEDIUM TANKS M3, M3A1, AND M3A2

TM 9-750
125

b. Disconnect the flarenut on the oil return line at the piston. Loosen, but do not remove the ffarenuts at the upper and lower connections of the cylinder. This will permit venting the air from the oil lines and from the cylinder.

c. Remove the cap from the oil level cup.

d. Fill the pump and lines with oil through the oil reservoir until an oil drippage is noticed at the upper and lower connections of the cylinder, and the oil flows freely from the oil return line.

NOTE.—The oil should be at approximately 70° F. temperature. If necessary, heat the oil before charging the system.

e. Reconnect the oil return line and tighten the flarenut permanently.

f. Take up the upper cylinder flarenut sufficiently to prevent leakage, but do not tighten permanently.

g. With the hand elevating gear disengaged, slowly push the breech of the gun all the way down to the end of the piston stroke. This action pushes the air out of the lower side of the cylinder.

h. With the breech being held in the lowest position, tighten the lower flarenut snugly and then loosen the upper cylinder flarenut (⅜ inch).

i. Slowly move the breech to the uppermost position, raising the piston to the top of the stroke, and tighten the upper cylinder flarenut (⅜ inch) snugly. This action pushes the air out of the upper side of the cylinder.

j. Loosen the bottom flarenut and repeat steps *g*, *h*, and *i* above. Tighten both flarenuts permanently.

k. Disconnect the multi-prong connector from the control unit.

l. Throw the master switch to the "on" position.

m. Operate oil pump from 15 to 30 minutes.

Caution.—Keep the oil reserve two-thirds full throughout this operating period.

n. Throw the master switch to the "off" position.

o. Slowly pump the breech of the gun up and down until no more bubbles appear in the oil reservoir.

p. Replace the oil reservoir cap.

q. Reconnect the multi-prong connector to the gyro-control unit.

r. Start the stabilizer equipment and check operation.

NOTE.—At the 25-hour check the oil system must be checked for presence of air by repeating step *o* above. Repeat the same step at any time to check for air in the system as a cause of erratic operation.

135

TM 9–750

ORDNANCE MAINTENANCE

126. Replacement of assemblies.—*a. Control unit.*—Either control unit is removed as follows:

(1) Place the master switch in the "off" position.

(2) Remove the multi-prong electrical connector from the control base.

(3) Remove the nuts on the control mount shaft and remove the control unit.

NOTE.—The lead seals on the control unit must *never* be broken except by ordnance personnel. The control is never repaired, but always replaced as a unit.

(4) Attach the control unit in the reverse order of removal.

(5) After replacement, check for proper operation.

b. Piston and cylinder.—(1) Throw the master toggle switch to the "off" position, disconnecting the power supply to the stabilizer.

(2) Disconnect and remove the oil lines from the cylinder assembly and plug with flareplugs.

(3) Loosen the clamping bolts (fig. 79) and remove the cylinder pivot pin at both ends of the assembly.

(4) Remove the piston and cylinder assembly.

(5) Replace in the reverse order of removal.

(6) Check to see that the piston and cylinder assembly is functioning freely without binding.

c. Repair and replacement of defective oil lines.—(1) Disconnect the defective oil line and catch any oil drainage from the affected part of the system in a container.

(2) Reflare the tubing using a standard flaring tool (fig. 81). The tubing should be cut as close as possible to the flare with a tubing cutter and *not* with a hacksaw.

NOTE.—Place flarenut on the tubing before the flare is made.

(3) Reconnect the flarenut to its proper connection.

(4) Replace completely any damaged oil lines.

(5) If connecting elbows cannot be tightened sufficiently to stop leakage, remove and tin threads with soft solder.

(6) Recharge the system with oil as outlined under paragraph 125.

d. Replacing shielded cable and wires.—(1) If the shielded cable connecting any part of the stabilizers becomes frayed or broken it should be replaced.

(2) Disconnect all wires running in the defective shielded cable.

(3) Disconnect and remove the defective shielded cable and wiring.

(4) Measure a new piece of shielded cable to the proper length.

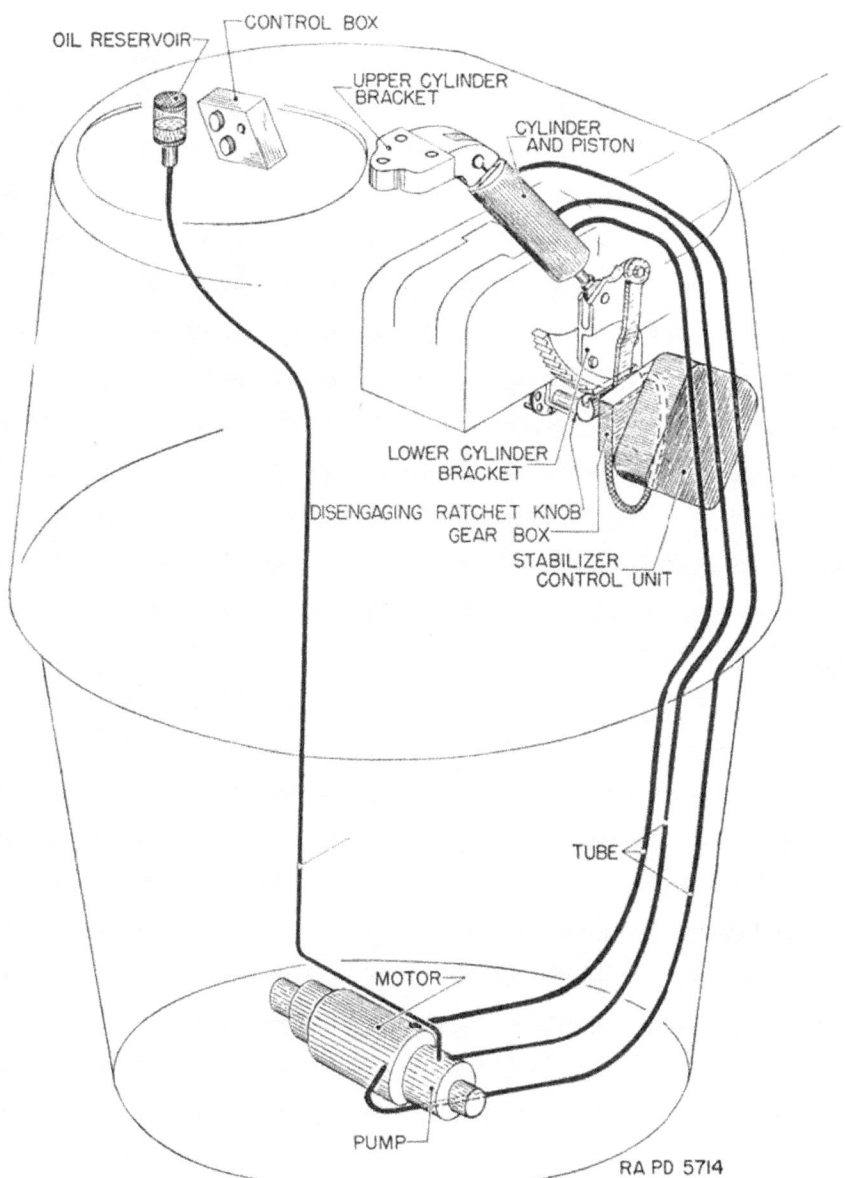

Figure 76.—37-mm gun stabilizer system.

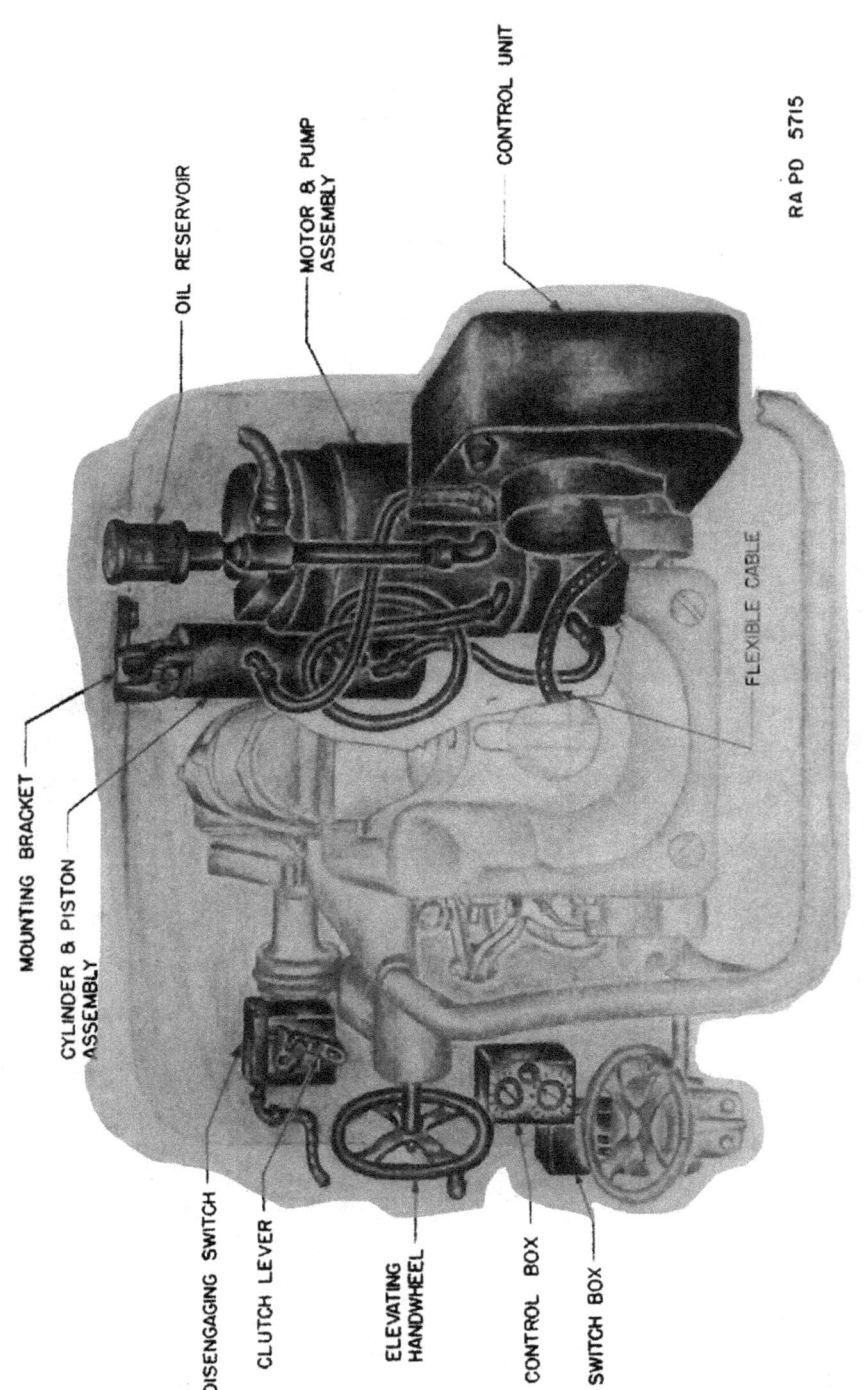

Figure 77.—75-mm gun stabilizer system.

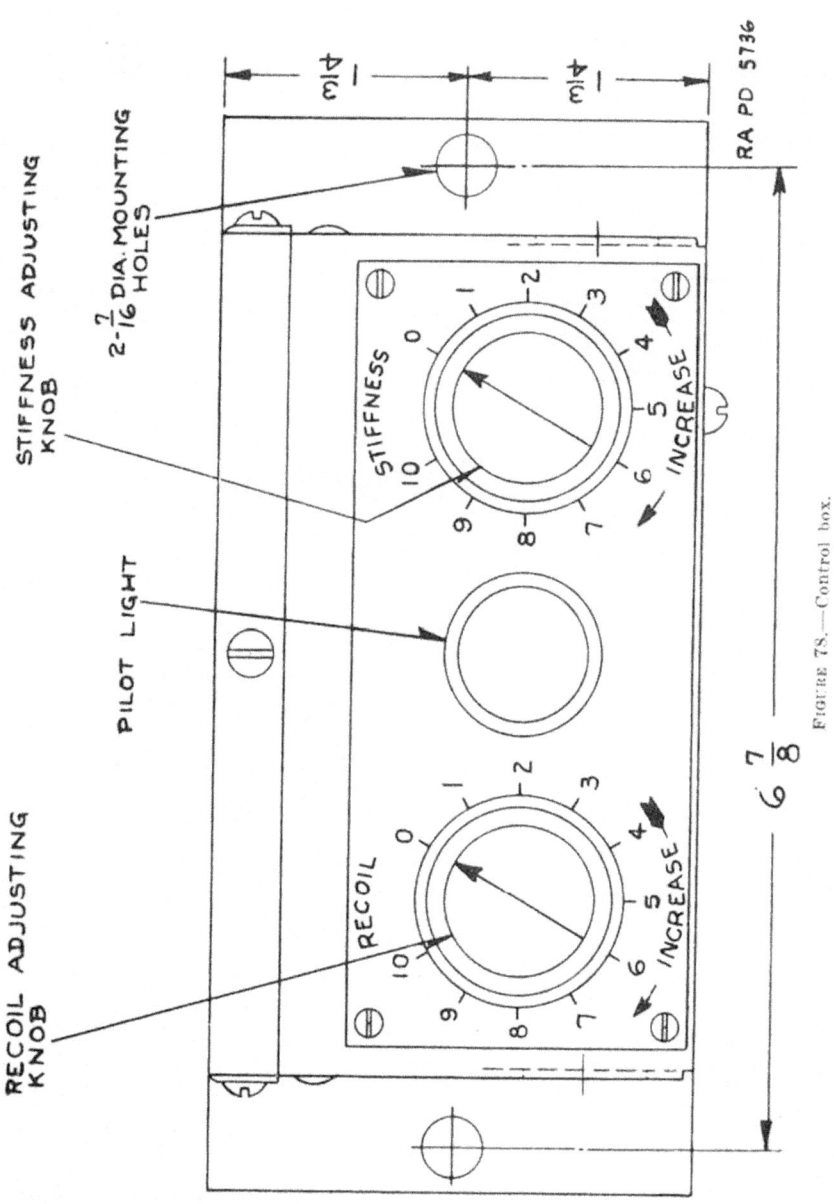

Figure 78.—Control box.

TM 9-750

ORDNANCE MAINTENANCE

(5) Tin the new shielded cable at least one-half inch on each side of where the cut is to be made.

NOTE.—Use a noncorrosive flux.

(6) Hold the cable securely and cut as measured, using a fine-toothed hacksaw.

(7) Tin the inside of a cable adapter fitting.

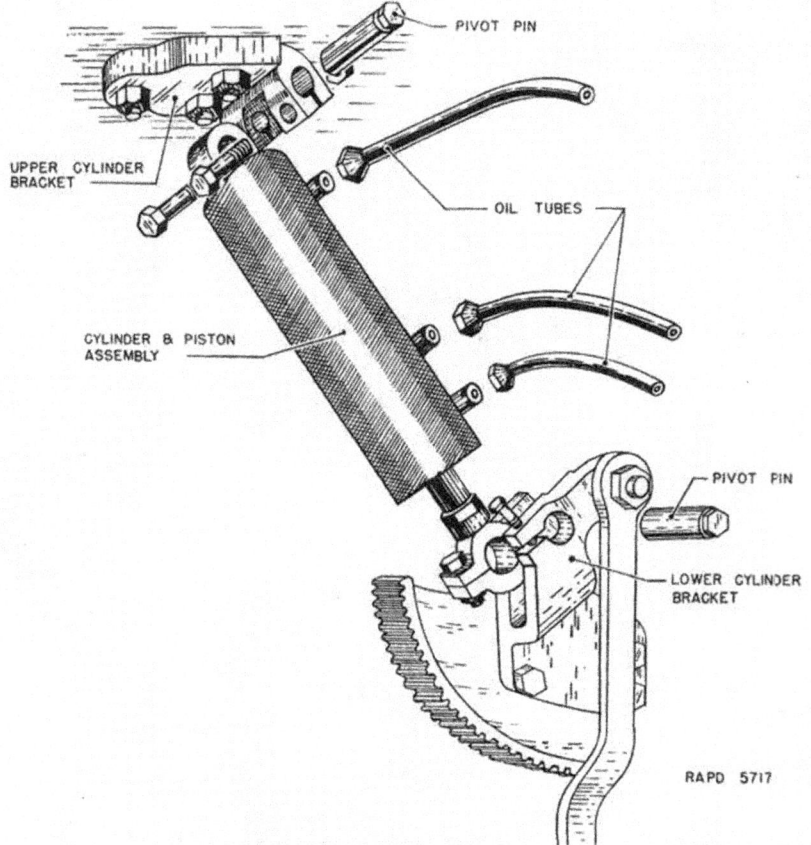

FIGURE 79.—Piston and cylinder.

(8) Insert the cable in the fitting and sweat solder it in place.

NOTE.—The torch flame must be on the fitting only, and never directly on the cable itself.

(9) Measure and cut to length new wire of the correct size and color.

(10) Run the wires through the shielded cable.

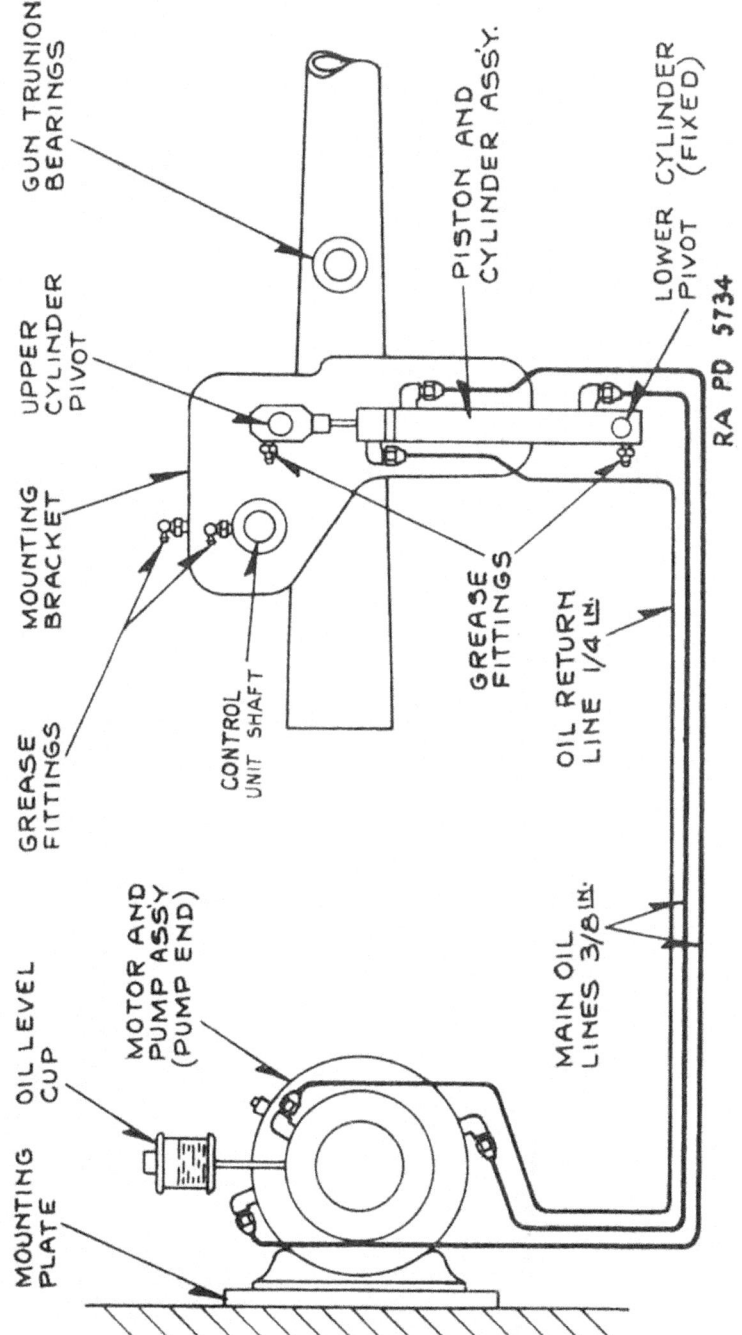

FIGURE 80.—Lubrication system.

TM 9-750

ORDNANCE MAINTENANCE

(11) Reconnect all wiring by colors to the same connections from which the damaged ones were removed.

e. Replacing grease fittings.—(1) The grease fittings are of the alemite type and are located as follows (fig. 80):

(*a*) Two on mounting bracket, greasing the control mounting shaft.

(*b*) One on lower side of the piston, greasing the cylinder pivot pin.

(*c*) One on piston rod end, greasing the piston rod end pivot pin.

(2) These fittings may be replaced by unscrewing and replacing with new ones.

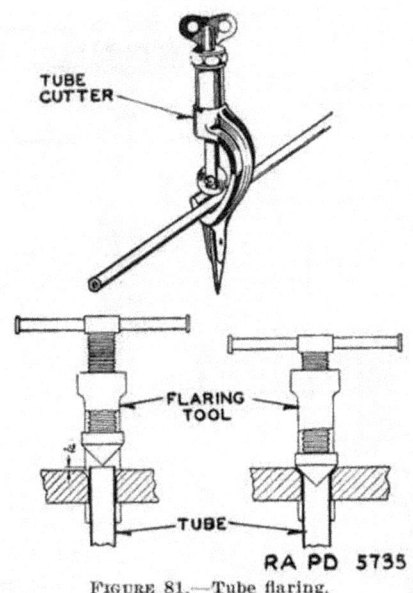

FIGURE 81.—Tube flaring.

SECTION XIV

TURRET

	Paragraph
General description	127
Operation of hydraulic turret control	128
Electrical system	129
Collector ring	130
Cupola	131

127. General description.—The turret is a one-piece casting of 2-inch armor which rotates on a ball-bearing race recessed and protected against direct hits and lead splash from enemy fire. The turret shield or basket is rigidly fastened to the cast turret, which projects above the roof of the tank, by means of a ring of bolts around its top circumference.

128. Operation of hydraulic turret control (fig. 82).—*a.* The turret is traversed by either the hydraulic or the manual traversing systems.

b. To operate hydraulic traversing system, first close the motor switch by pressing down on the knob. Pressure in the hydraulic system will build up almost instantly. For hydraulic operation, the clutch lever should be in the "up" position. To traverse the turret, grasp the pistol-grip control, squeeze the trigger, and turn the control handle. Counterclockwise turning of the control handle revolves the turret to the left. Reversing the direction of the control handle reverses the direction of rotation. Complete 360° rotation is obtainable. The farther the handle is turned in either direction from its neutral position the faster the turret moves.

Caution.—Be sure that the position lock (which is a left-hand thread) is not engaged before any attempt is made to traverse turret.

c. If the power supply or the hydraulic mechanism fails, the turret may be traversed by throwing the clutch lever in the "down" position and operating the hand crank. Fluid present in the lines acts as a brake preventing any movement of the turret when the control valve is in neutral. When the hand mechanism is being used, an automotive type of brake with Wellman lining incorporated in the hand control acts to prevent undesired movement of the turret. The hydraulic traversing system consists essentially of three units: the pump, the control-valve mechanism, and the hydraulic motor. The pump, directly connected to an electric motor, is mounted on the floor of the basket under the 37-mm gunner's seat. The control valve mechanism is mounted to the left of the 37-mm gun mount at the same height as the traversing race. Turning of the pistol-grip control handle discharges oil to one of the two sides of the hydraulic motor and traverses the turret as previously described. The hydraulic motor and mechanism are mounted slightly to the left of the control mechanism. The hydraulic motor is of a reversible-gear type, directly connected to a gear train. A pinion engages the traversing gear surrounding the turret. The manual traversing mechanism is mounted on the housing of this unit and a lever projecting from the housing is provided for changing from hydraulic-powered traversing to manual operation. The connections between these units and to the oil reservoir are by high-pressure rubber hose and copper tubing. The reservoir mounted on the floor of the basket contains approximately 2.2 gallons of a special oil (see sec. **IV**, ch. 1).

TM 9-750
ORDNANCE MAINTENANCE

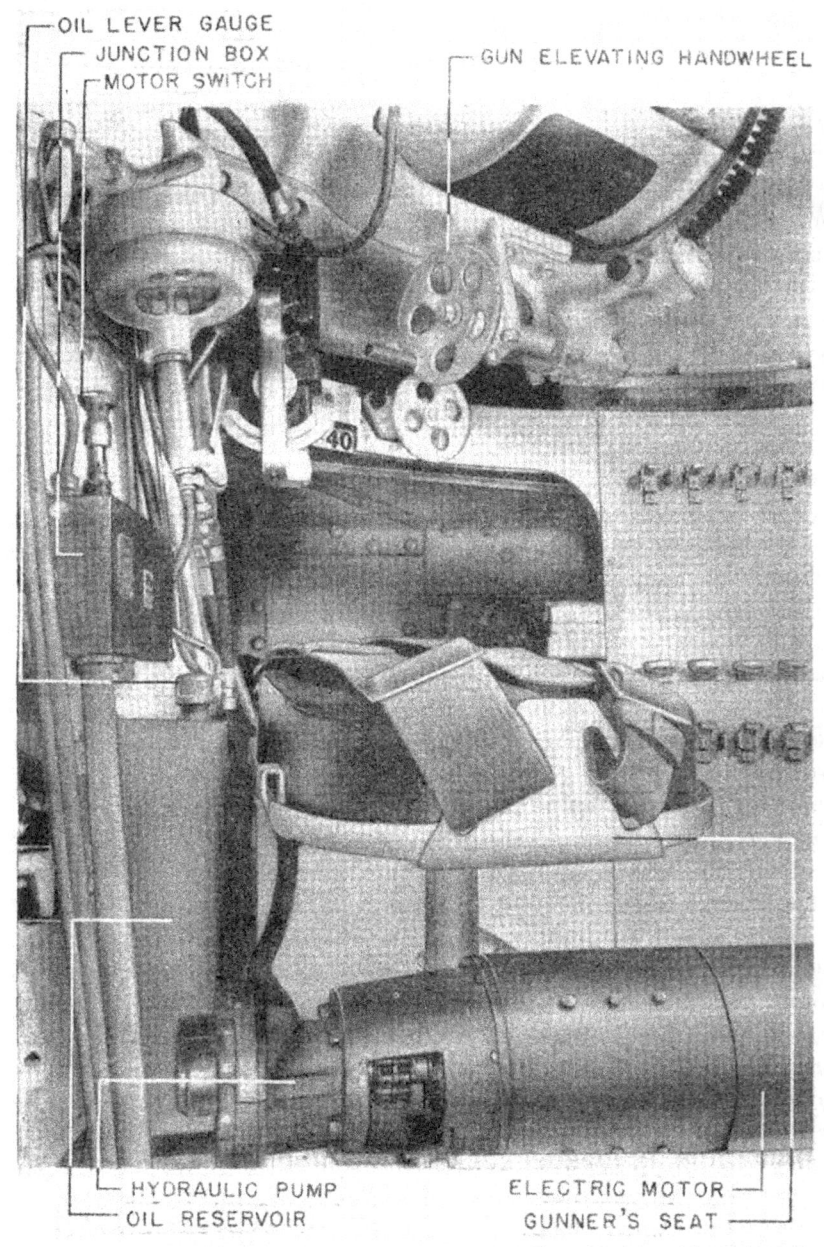

FIGURE 82.—Interior view of turret.

The oil level should be maintained at all times and the mechanism should not be operated without oil, since it acts also as a lubricant as well as a power-transmitting medium. In an emergency, SAE 20 lubricating oil may be used; however, it should be replaced with the proper oil at the earliest opportunity. Incorporated in the hydraulic control valve is a relief valve which determines the system pressure. A breather cap takes care of the expansion of the oil. Attached to the handle of the operating valve are two toggle switches which

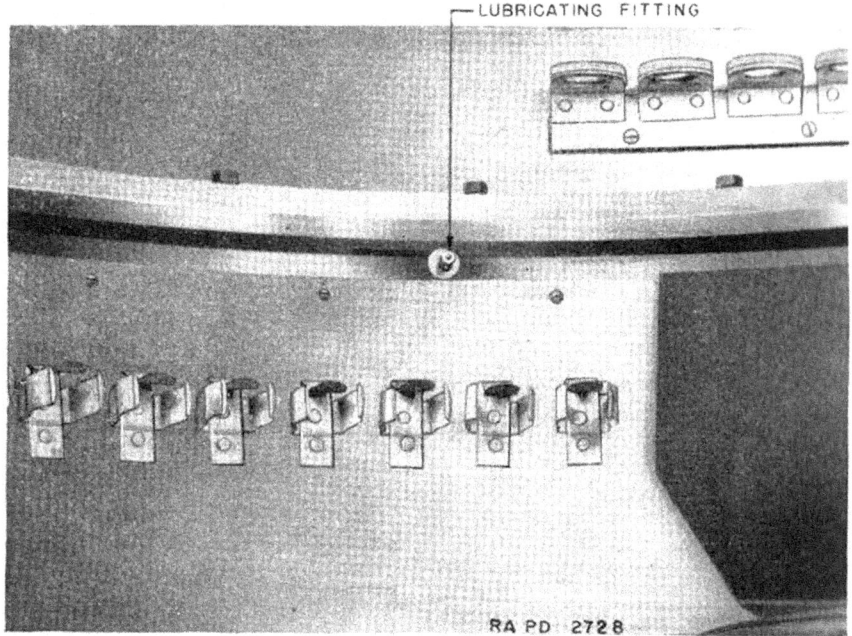

FIGURE 83. One of the lubricating fittings for turret race.

control the firing of the 37-mm gun and caliber .30 machine gun. It is necessary to squeeze the pistol-grip handle firmly before tripping the firing switches in order to release the safety mechanism. One of the lubricating fittings for the turret race are shown in figure 83.

129. Electrical system (figs. 68 and 69).—Electrical power is transferred from the battery in the hull section to the turret by means of a collector ring assembly. The guns in the turret are fired from the switches on the hydraulic control handle by solenoids which operate the firing levers. A safety switch is incorporated in the operating handle. The power lead from the collector ring assembly goes through the motor switch and electric motor. The interphone system take-off leads come off the collector ring to a junction box.

130. Collector ring (fig. 84).—*a.* The purpose of the collector ring assembly is to transfer the electrical power from the main section of the tank to the turret. It consists of a cylinder which rotates with the turret and in sets of fixed brushes. The two top sets of brushes are power leads of which the top one is grounded. The bottom four are for the interphone communication system.

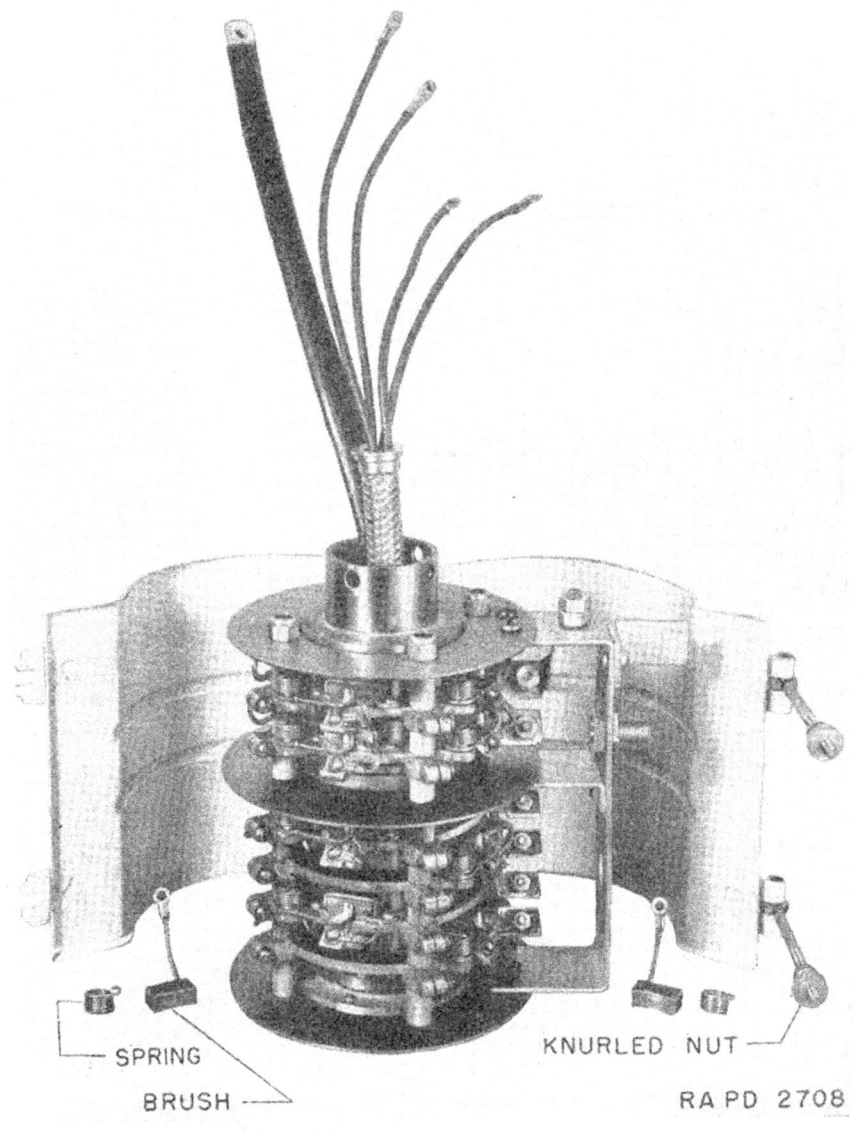

FIGURE 84.—Collector ring assembly.

b. There is little maintenance required on this unit. With the main battery switch open, remove the cover by loosening the two knurled nuts. Check to see if the brushes are free in the brush boxes. Brushes can be replaced by removing the screw from the brush lead.

131. Cupola.—Lubrication fittings for the cupola are shown in figure 85.

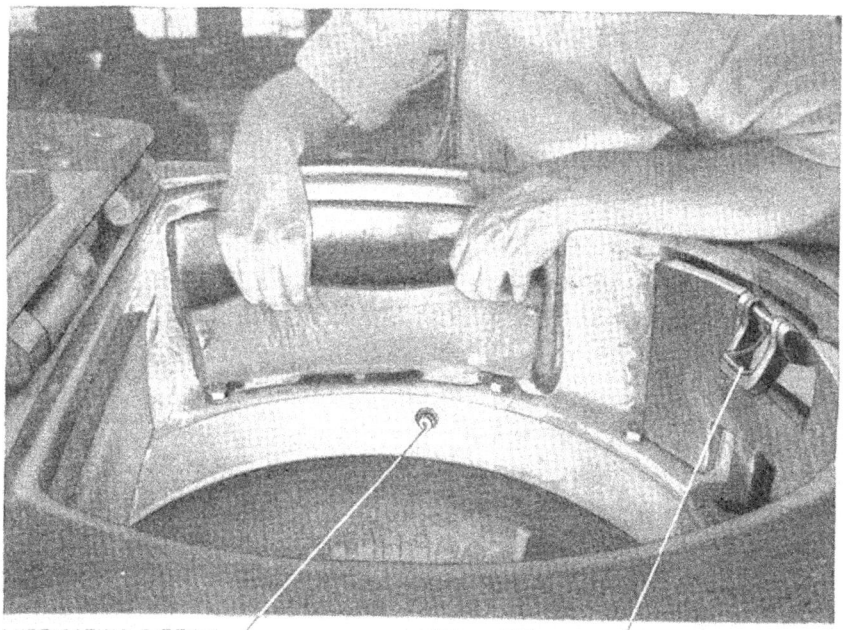

FIGURE 85.—Lubrication fitting in cupola.

Section XV

PREPARATION FOR SHIPMENT AND STORAGE

	Paragraph
Shipment	132
Storage	133
Storage of components and equipment	134

132. Shipment.—When shipping by rail on flat cars, every precaution must be taken to have the tank properly fastened and blocked to the floor of the car.

a. Cars must be inspected to see if they are in suitable condition to carry the load to its destination safely. The weight of the tank, approximately 30 tons, makes it most necessary that the cars have good, sound floors. All loose nails or other projections not an integral part of the car, other than the prescribed blocking, must be removed. Loose nails, bolts, etc., necessary in car construction, should be made tight rather than removed.

b. The load must be so distributed that there will not be more weight on one side of the car than on the other. If a single tank is placed on a car it must be so located that one truck of the carrying car does not carry more than one-half of the load weight. All doors and other openings should be closed and securely fastened as a protection against weather and pilfering of equipment. Material used for blocking must be of hardwood, fir, spruce, or long-leaf yellow pine, straight grain and free from strength impairing knots.

c. If the vehicle is to be used immediately upon reaching its destination, it may be desirable to keep the fuel tanks filled. If transportation of the tank is by rail express, in which case civilian passengers may be carried in coaches of the same train, the fuel tanks must be drained. The draining of the fuel tanks is not required if only military personnel is carried in accompanying coaches.

d. Equipment moving from manufacturer to arsenal or proving ground, or from arsenal or proving ground to Army post, or individual units moving from one Army post to another *must* be placarded "DO NOT HUMP."

e. Further details on loading and shipping are to be found in "Special Supplement Containing Rules Governing the Loading of Mechanized and Motorized Army Equipment." Another source of information is "Major Caliber Guns For the United States Army and Navy, on Open Top Equipment" published by the Association of American Railroads, Operations and Maintenance Department, April 1, 1941. FM 101–10 also gives information on shipping instructions.

f. Before fastening the tank to the car it will be necessary to take certain measures to prevent corrosion during transit. If the vehicle is to be in transit for a period of less than 30 days it will be permissible to prepare it by operating the engine for a period of not less than 1 hour, on unleaded "white" gasoline. However, if it is at all practicable, spraying of the interior of the engine with aircraft engine corrosion-preventive compound, according to the methods outlined in AR 850–18 is desirable. In the tank this will not be possible without first removing the engine. If the vehicle is to be put in dead storage after it reaches its destination, the engine should be removed and shipped separately in order that corrosion-preventive treatment can be applied immediately before shipment and shortly after it reaches its destination and is placed in dead storage. Due to the corrosive nature of the high octane gasoline ordinarily used in the tank engine it is most necessary that the above-mentioned corrosion-preventive measures be taken. Failure to do so will result in serious damage by corrosion. Paragraph 133 outlines the methods of applying corrosion-preventive compounds.

133. Storage.—*a. General instructions.*—The storage of motor vehicles and equipment and inspection in connection therewith are covered in AR 850–18. With a few exceptions, the information given in this manual will apply directly to storage and the preparation for storage of the medium tanks M3, M3A1, and M3A2. It will not be practicable in many cases to apply the internal corrosion-preventive treatment to the engine prior to limited storage because of the difficulty in removing spark plugs and spraying the interiors of cylinders and other parts while the engine is in the tank. Insofar as it is possible to do so, methods outlined in AR 850–18 must be followed in order that engines and other portions of the matériel are properly protected from damage by corrosion. For dead storage, engines must be removed from the tank, inspected, reconditioned if necessary, and then given the proper corrosion-preventive treatment and stored separately.

b. Storage conditions.—Tanks which are not in actual use will be stored in closed buildings or covered sheds if available. In lieu of this preferred storage space, they may be stored in the open and covered with tarpaulins. In each case, the floor must be solid and free from crushed rock, deep dust, and oil surfacing. It is desirable that the rubber tracks rest on planks. Every precaution must be taken to afford proper drainage of water from the floor and to locate the place of storage so that the matériel will be properly protected from flood or fire.

c. Technical inspections.—The tanks and equipment will be inpected at the time they are placed in storage and at frequent periods as designated in AR 850–18. A tag or tags tied to the unit or vehicle will be kept up to date by the inspector indicating the condition and work to be done before the unit is again placed in service. Minor work of surface preservation and application of corrosion-preventives will be accomplished at the time of inspection. Work involving the use of shop facilities will be accomplished at the earliest practicable date. Batteries should be removed when a vehicle is placed in dead storage and the battery placed in active stock if practicable.

d. Limited storage.—Vehicles in limited storage are those that are out of service for less than 30 days. Vehicles stored under this heading will be ready for immediate service, and the fuel tanks and oil tanks will be kept filled. The battery will be maintained in a

TM 9-750

ORDNANCE MAINTENANCE

fully charged condition and should remain in the tank. The battery switch and radio switch will be open during storage periods. The vehicle must be thoroughly cleaned and lubricated before being placed in limited storage, and proper precautions should be taken to protect the rubber elements from extreme light or heat. Breaks will not be left applied and the vehicle will not be left in gear.

e. Placing vehicles in dead storage.—Vehicles in dead storage are those that will not be required for service for an indefinite period. Vehicles should not be in limited storage status for over 30 days; however, it will be impossible to adhere rigidly to this ruling under certain combat conditions. When the medium tank is placed in dead storage, the engine will be removed, inspected, repaired at once if practicable, and treated with corrosion-preventive compounds. In any event, it is most essential that the treatment with corrosion-preventive compounds be given immediately after the engine is placed in dead storage.

(1) *Engine, valve compartment.*—Remove the rocker box covers and spray the valve mechanism with aircraft-engine corrossion-preventive compound. Special spraying outfits are available for applying this compound. The engine crankshaft will be rotated while the corrosion-preventive compound is applied in such manner as to cover thoroughly the entire surface of cams and protruding ends of valve stems.

(2) *Cylinders.*—The cylinder walls, piston heads, and valves will be treated with aircraft-engine, corrosion-preventive compound. Set the engine with the crankshaft extending vertically upward and the piston of the cylinder which is being sprayed placed at bottom center of the suction stroke. Spray each cylinder through the spark plug or injector hole, taking care not to damage the threads during the spraying operation. The quantity of compound to be sprayed into each cylinder will be metered, if possible; if meters are not available, measurement will be made by determining accurately the time required to spray the specific quantities and timing subsequent spraying of the individual cylinders to obtain the proper quantity in each. In first performing the spraying operation, determine by experiment the proper technique for completely coating cylinder wall, piston head, and valves with corrosion-preventive compound. Spray 2 ounces ($\frac{1}{8}$ pint) of compound into each cylinder, then turn the engine crankshaft at least two complete revolutions. Respray the cylinder space above each piston with $\frac{1}{16}$ pint (1 ounce) of compound without revolving the crankshaft further, since this space is particularly susceptible to corrosion. This procedure permits

treatment of the maximum surface of the cylinder wall and the piston head. The intake valve will be open sufficiently to permit a small quantity of the vaporized corrosion-preventive compound to reach the valve face. After thus treating the interior of all cylinders, spray a small quantity of the corrosion-preventive compound into each exhaust port, with each exhaust valve in a fully open position so that it will be coated. If the engine to be treated is equipped with exhaust collector rings that are difficult to remove, spray the exhaust valve through the spark plug hole instead of the exhaust port. However, this must be done with the exhaust valve fully open and after the inside of the cylinder has been coated as outlined above.

(3) *Magnetos.*—Magnetos will be treated by lightly coating the steel parts with rust-preventive compound, or with petrolatum.

(4) *Openings.*—Close all fuel and oil lines or open connections, cylinder ports, and other openings with suitable plugs, covers, etc. Close threaded openings with threaded plugs whenever practical. If wooden or similar tapered plugs are used, they must be so constructed that they cannot be accidentally pushed or driven completely into the opening. When necessary, remove the rear oil tank vent connection plug to prevent excess oil in the rear section from draining into the magneto while the engine is standing with the crankshaft in the vertical position. When this plug is removed, wire it to the engine as close as possible to the hole from which it was removed.

(5) *Spark plugs.*—Remove spark plugs, oil the spark plug holes, and then stop them by inserting shipping plugs, unserviceable spark plugs, corks, or wooden plugs. Clean, adjust, and test the removed plugs; if they are serviceable, coat them with corrosion-preventive compound and place them in stock.

(6) *Exterior of engine.*—Clean the exterior of the engine thoroughly with dry-cleaning solvent. Apply a coating of rust-preventive compound on all unpainted steel parts. Remove rust appearing on any part before storage with sandpaper or a wire brush and coat the metal with corrosion-preventive compound unless the surface was originally painted. Painted surfaces will be repainted.

(7) *Gasoline.*—Drain gasoline and return it to proper storage.

(8) *Battery.*—Remove the battery from the vehicle. After plugging the vents in the cells, clean the case with a solution of soda ash and water (8 ounces to the gallon) to neutralize the acid. After this treatment, flush the case with cold water; do not use hot water or steam. Remove plugs from the vents after cleaning. Clean terminals and cable ends thoroughly with soda ash solution, scraping them clean with a suitable tool or wire brush and then grease with petrola-

tum or light grease. Take hydrometer readings of the cells, and charge the battery if readings are 1.225 or less. Add distilled water to cover the plates, but not more than one-quarter inch above plates. Place the battery in active stock. Never allow batteries in stock to become discharged below a hydrometer reading of 1.225; this will be a proper precaution against freezing in all but the most severe weather, when a specific gravity of 1.250 must be maintained.

(9) *Vehicle general.*—Remove rust appearing on any part before storage with sandpaper. Painted surfaces will be repainted and unpainted surfaces will be lightly coated with rust-preventive compound.

(10) *Inspection tag.*—Attach a tag to the dashboard. The dates of inspections will be entered on this tag and each initialed by the inspector.

f. Periodic treatment of vehicles in dead storage.—At the expiration of each 3-month period, repeat corrosion-preventive treatment, giving particular attention to the cylinders, valve compartments, and other internal parts. Under unfavorable climatic conditions such as might occur in tropical climates or near salt water it will be necessary to perform more frequent inspections and corrosion-preventive treatments in order to prevent damage to the equipment.

g. Check-up of vehicles in dead storage.—Vehicles will be inspected frequently to see that equipment or parts are not removed without proper authority.

h. Removing vehicles from dead storage.—(1) *Magnetos.*—Wipe the magnetos with a clean dry rag to remove excess corrosion-preventive compound used and assemble magneto to the engine.

(2) *Cylinders.*—Remove plugs from spark plug holes and pump out excess corrosion-preventive compound above the pistons with a hand pump. If no pump is available, turn the engine over by hand to force out the corrosion-preventive compound.

(3) *Valves.*—Rotate the crankshaft through three or four revolutions by hand and observe for proper operation of valve mechanism. Lubricate the stem of any valve found to be sticking with penetrating oil or with a 50-50 mixture of kerosene and light lubricating oil. Continue to turn the engine over by hand until all evidence of sticking valves has been eliminated. If this treatment does not free the valves, necessary mechanical repairs to free them must be made before the engine is placed in service.

(4) *Gasoline tank.*—Fill the tank.
(5) *Spark plugs.*—Install new spark plugs in the engine.
(6) *Battery.*—Install a fully charged battery.

(7) *Transmissions, transfer cases, differentials, axle housings. etc.*—Drain old lubricant from these units and other inclosed gears, flush thoroughly with lubricating oil SAE 10, and fill them to proper levels with the correct lubricant, using the lubrication section as a guide.

(8) *Engine oil tank.*—Install the engine in the tank, and fill the oil tank about half full (19 qt.). Since the material used as a corrosion-preventive on the interior surfaces of the engine mixes freely with engine oil, it will not be necessary to remove it prior to adding lubricating oil. Run the engine 2 or 3 hours, then drain the oil tank and add new oil to the system. The second addition of oil will require 36 quarts to fill the oil tank properly.

(9) *Lubrication.*—Lubricate the tank thoroughly before it is placed in service.

(10) *Inspection.*—A thorough inspection of the vehicle will be made upon removal from dead storage. Any repairs that have been ordered on the inspection tag attached when the vehicle entered storage and which have not been previously made must be taken care of at this time.

(11) *Starting engine.*—Start the engine according to the instructions given in section III, chapter 1. Particular attention should be given to watching for overheating of the engine, excessive vibration, or any unusual noises that may indicate something is wrong. Starting of the tank by coasting or by towing must not be attempted, since serious injury to the working parts may result from this practice.

134. Storage of components and equipment.—Components removed from vehicles prior to storage must be thoroughly inspected and overhauled if necessary before being stored for reissue. These components, including engines, need not be retained in storage for any particular vehicle, but should be considered as stock when issue becomes necessary.

Appendix

LIST OF REFERENCES

1. Standard Nomenclature Lists.

Gun, machine, cal. .30, Browning, M1919A4, fixed and flexible, and mount, tripod, M2	SNL A-6
Gun, submachine, cal. .45, Thompson, M1928A1	SNL A-32
Cleaning and preserving materials	SNL K-1
Tank, medium, M3 (in preparation)	SNL G-104

2. Field and Technical Manuals.—*a. General.*

Cleaning and preserving materials	TM 9-850
Automotive lubrication	TM 10-540
Automotive electricity	TM 10-580
Motor transport	FM 25-10
Tank units	FM 7-10
Defense against chemical attack	FM 21-40

b. Armament.—The following publications describe in more detail the armament of the tanks:

37-mm gun, M5 (mounted in tanks)	FM 23-80
37-mm gun, M6 (mounted in tanks)	FM 23-81
Browning machine gun cal. .30, HB M1919A4 (mounted in combat vehicles)	FM 23-50
Thompson submachine gun, cal. .45, M1928A1	FM 23-40
75-mm gun matériel, tank	TM 9-307
Gun, 37-mm, M5 and M6	SNL A-45
Matériel, 75-mm tank gun, M2A1	SNL C-34

TM 9-750

INDEX

	Paragraph	Page
Accessories	42–60	58
Air cleaners	56	69
Ammunition, decontamination	34	33
Ammunition stowage	35	35
Auxiliary generating unit:		
Description	120	127
Maintenance	121	128
Battery	101	106
Bogie	96	95
Bogie, wearing plate	96	95
Bogie wheel	96	95
Booster coil	51	66
Bulkhead, engine	61	76
Carburetor	55	68
Chemical attack, cleaning after	34	33
Cleaning	26	30
Clutch	70	82
Clutch adjustment	74	84
Cold weather starting and operation	14	16
Collector ring	130	146
Control, voltage	103	108
Controls, operator's	6	10
Cooling system, engine	68	81
Cupola	131	147
Decontamination	34	33
Description, general	4	2
Differential	85	88
Draining fuel tanks	61	76
Driving instructions	11	14
Engine:		
Characteristics	44	60
Descriptive data	42	58
Hoisting	59	70
Installation	59	70
Lubrication	18, 46	21, 60
Mounting	59	70
Removal	59	70
Equipment supplied	40	52
Filter, generator	104	111
Final drive	91	93
Fire extinguishers	41	53
Fuel pump	54	68

TM 9-750

INDEX

	Paragraph	Page
Fuel, shut-off valves	62	78
Fuel tanks	61	76
Fuse box	115	119
Gages, fuel	65	80
Gasoline, grade	67	81
Gasoline tanks	61	76
Gear shift lever	89	90
Generator	53	67
Governor	57	70
Grousers	99	106
Guns	35	35
Hub assembly	95	94
Idler	98	102
Ignition harness	48	61
Indirect vision windows	37	45
Inspection:		
50-hour	24	28
100-hour	24	28
After operation	23	27
Before operation	7, 20	11, 25
During operation	21	26
Installation of engine	59	70
Instrument panel	8, 106	12, 113
Instruments	106	113
Interphone	118	120
Lights	113	117
Lubrication chart	18	21
Lubrication, engine	16, 46	17, 60
Magneto	49	63
Manifold, engine	47	61
Maintenance	38	47
Muffler	47	61
Oil dilution valve	17	19
Oil filter	16	17
Painting	27	31
Pedal, clutch	74	84
Pistol port	35	35
Plates, clutch	80	86
Primer, engine	63	79
Propeller shaft	82	88
Protectoscope	37	45
Pump:		
Engine oil	46	60
Primer	63	79
Push rod, replacement	60	73
Radio	117	120
Records	25	30

INDEX

	Paragraph	Page
Removal of engine	59	70
Shielding, radio	116	120
Shipment	132	147
Sight, gun:		
75-mm	35	35
37-mm	37	45
Siren	114	119
Solenoids, firing	110	116
Spark plugs	50	64
Sprockets	95	94
Stabilizer	36, 122–126	44, 133
Starter, electric	52	67
Starting instructions	8	12
Steering brake adjustment	87	89
Storage	134	153
Strainer, fuel	64	80
Submachine guns	35	35
Switch:		
Battery	102	108
Starter	105	113
Tabulated data	5	7
Track	97	99
Transmission	85	88
Tripods	35	35
Trouble shooting, engine	39	50
Turret:		
Description	127	142
Electrical system	129	145
Hydraulic control	128	143
Valve push rod, replacement	60	73
Valve rocker arm, replacement	60	73
Valves:		
Adjustment	60	73
Fuel, shut-off	62	78
Volute springs	96	95

[A. G. 062.11 (8–4–41).]

BY ORDER OF THE SECRETARY OF WAR:

G. C. MARSHALL,
Chief of Staff.

OFFICIAL:
    J. A. ULIO,
        Major General,
            The Adjutant General.

DISTRIBUTION:
    D 17 (10); R 17 (6); IBn 17 (4).9 (1); IC 17 (17), 9 (3).
    (For explanation of symbols see FM 21–6.)

Also Now Available!

Visit us at:

www.PeriscopeFilm.com

©2013 Periscope Film LLC
All Rights Reserved
ISBN#978-1-937684-35-8
www.PeriscopeFilm.com

www.ingramcontent.com/pod-product-compliance
Lightning Source LLC
Chambersburg PA
CBHW071718090426
42738CB00009B/1812